THE FUNDAMENTAL THEOREM OF ARITHMETIC

With an elementary introduction to prime numbers

Ευκλείδης:

Πρώτος αριθμός εστίν ο μοναδι μόνη μετρούμενος

Euclid:

A prime number is a number measured by a unit alone

The book contains 60 solved illustrative exercises and 115 problems for solution. Odd numbered problems are provided with answers. Hints or detailed outlines are given for the more involved problems.

The Fundamental theorem of arithmetic (with an elementary introduction to prime numbers)

Demetrios P. Kanoussis, Ph.D

PREFACE

A prime number is a natural number greater than 1, which cannot be expressed as a product of two smaller natural numbers. Stated differently, a prime number is a number divided only by 1 and by itself. No other natural number can divide a prime number. For example, the numbers: 2, 3, 5, 7, 11, 13, 17, 19, etc, are prime numbers. All the other numbers are called composite numbers. For example, the numbers: 4, 6, 8, 9, 10, 12, 14, 15, 16, 18, 20, etc, are composite numbers. The prime number 7 is divided by 1 and 7 only, while the composite number 12 is divided by 1, by 12, but also, by 2, and by 3, and by 4, and by 6.

Prime numbers, despite their simple definition, still hide many secrets, and their intensive study has been the subject of many eminent mathematicians, for more than 25 centuries. Prime numbers and their properties were studied, for the first time, by the ancient Greeks. Euclid defined a prime number as "**a number measured by a unit alone**" and also gave the first proof that there are infinitely many prime numbers, (around 300 BC). Eratosthenes of Cyrene, around 200 BC, devised a procedure, (an algorithm), to find all the prime numbers less than a given natural number, (the sieve of Eratosthenes).

The main problem with prime numbers is that the way they appear is highly irregular. There is no known formula, or method, to predict the next prime number from the preceding one. Great mathematicians, Gauss, Euler and Riemann among others, have made important contributions to the prime numbers, but, the main problem of finding a pattern as how the prime numbers are distributed, still, remains unsolved. There are also, many other unsolved problems concerning the prime numbers. One of them is the **Goldbach's conjecture**, that every even number greater than 2 can be expressed as the sum of two primes. Another famous unsolved problem is **the twin prime conjecture**, that there are infinitely many pairs of primes that differ by 2, (like 3 and 5, 5 and 7, 11 and 13, 17 and 19, etc). Other unsolved problems concerning prime numbers and some historical remarks, like, Euclid's theorem, the sieve of Eratosthenes of Cyrene, Fermat numbers, etc, are discussed in section 3-4.

Primes are of fundamental importance in number theory, because of the fundamental theorem of arithmetic: **every natural number greater than 1, can be expressed as a product of prime numbers, and this product representation is unique**. In this little book we provide a rigorous proof of this theorem and present some of its applications, (like for instance, to find the number of divisors and the sum of the divisors of any natural number, the perfect numbers and the amicable numbers).

The material covered in the book is shown analytically in the table of contents. A comprehensive introduction to the prime numbers and some of its more important properties are presented, which are essential for the proof of the fundamental theorem.

Solving problems with prime numbers is not an easy task. Most of the times, the solution requires ingenuity and imagination. Some other times, when a direct method of approach does not lead to the solution of the problem, we may proceed by contradiction, which is a popular and powerful method when working with prime numbers. The careful reader will find that many of the solved exercises have been solved by contradiction.

The book contains 60 illustrative, solved exercises and 115 problems for solution.

TABLE OF CONTENTS

CHAPTER 1: DIVISIBILITY, EUCLIDEAN ALGORITHM

1-1) Introduction

a) Arithmetic is considered to be the study of the properties of the integers and of the various operations with them. Contrary to the popular belief, arithmetic is one of the most difficult subjects in mathematics. This little book is intended to provide some elementary concepts, properties and theorems of the integers and build up the background needed in order to state and prove, rigorously, one of the most important theorems of arithmetic, the so called "**the fundamental theorem of arithmetic**".

Everyone is familiar with the whole numbers 1, 2, 3, 4, 5, 6, 7,….These numbers are also called **positive integers**, or **counting numbers**, or **natural numbers**, or even, **the integers of arithmetic**. The set of whole numbers is denoted by $\mathbb{N}$, i.e.

$$\mathbb{N} = \{1, 2, 3, 4, 5, 6, 7, \dots.\}$$

The first element in this set is the number 1. Every other element of the set is obtained from the preceding element by adding the number 1, i.e. 2=1+1, 3=2+1, 4=3+1, etc.

To denote that a number x is a whole number, we write $x \in \mathbb{N}$, (read: x belongs to $\mathbb{N}$). To denote that x is not a whole number, we write $x \notin \mathbb{N}$, (read: x does not belong to $\mathbb{N}$).

If we attach the number zero (0) to the set $\mathbb{N}$, we obtain the set $\mathbb{N}_0$, i.e.

$$\mathbb{N}_0 = \{0,1,2,3,4,5,6,7, \dots \}$$

The set of positive and negative integers, including the number 0, is called the **set of integers**, and is denoted by $\mathbb{Z}$, i.e.

$$\mathbb{Z} = \{\dots, -4, -3, -2, -1, 0, 1, 2, 3, 4, \dots \}$$

Sometimes, the set $\mathbb{Z}$ is called **the integers of algebra**.

The reader is supposed to be familiar with the elementary operations with integers, i.e. addition, subtraction, multiplication and division.

Note that, while addition, subtraction, and multiplication of integers yield an integer, (as we say, the set $\mathbb{Z}$ is closed under addition, subtraction and multiplication), the division of two integers, in general, is not an integer. For instance, the numbers: $5/3$, $17/9$, $-4/9$, etc. are not integers. These are **rational numbers**. The set of rational numbers is denoted by $\mathbb{Q}$. Obviously, $\mathbb{Z} \subset \mathbb{Q}$, (read: the set of integers is a subset of the set of rational numbers).

b) In what follows, we shall restrict our analysis to the set of numbers $\mathbb{N}_0 = \{0,1,2,3,4,5, ... \}$, i.e. to the set of **nonnegative integers**, since then, the extension of the results obtained to the negative integers is easy.

1-2) Multiples of integers

Multiples of an integer number a, are the products of a with the numbers $0, 1, 2, 3, 4, ...$, i.e. are the numbers $0, a, 2a, 3a, 4a, ...$

The general form of the multiples of an integer a, is ka, where $k = 0,1,2,3,4, ...$ Obviously, an integer has an infinite number of multiples.

If we know that a number c is a multiple of the integer a, we may write $c = ka$, $k \in \mathbb{N}_0$, or we may write that $c = mul. a$, (read: c is a multiple of a).

For instance, we may write: $8 = mul. 2$, (since $8 = 4 \times 2$), $24 = mul. 6$, (since $24 = 4 \times 6$), etc.

Properties:

1) The sum of two or more multiples of an integer a is another multiple of a.

Proof: Let, for instance, $c_1 = k_1 a, c_2 = k_2 a, c_3 = k_3 a$ be three multiples of the integer a, (recall that each one of the numbers k_1, k_2, k_3 are elements of the set $\mathbb{N}_0 = \{0,1,2,3,4, ... \}$). Then the sum $c_1 + c_2 + c_3$ shall be:

$$c_1 + c_2 + c_3 = k_1 a + k_2 a + k_3 a = \underbrace{(k_1 + k_2 + k_3)}_{integer} a = mul. a$$

since $(k_1 + k_2 + k_3)$ is an integer number.

For instance, since, $6 = mul. 3, 15 = mul. 3$ and $21 = mul. 3$, then, the number $(6 + 15 + 21) = 42 = mul. 3$, (indeed, $42 = 14 \times 3$).

2) The difference of two multiples of an integer a, is another multiple of a.

The proof is similar to the proof of property (1).

3) The multiple of a multiple of an integer a, is another multiple of a.

Proof: Let ka be a multiple of a. Then, any other multiple of (ka) will be of the form $\lambda(ka)$, where λ is an integer. But, $\lambda(ka) = (\lambda k)a = mul.\, a$.

For instance, since $21 = mul.\, 7$, then 63 which is a multiple of 21, $(63 = 3 \times 21)$, shall be also a multiple of 7. Indeed, $63 = 9 \times 7 = mul.\, 7$.

(The reader is supposed to be familiar with the elementary identities: $(a \pm b)^2 = a^2 \pm 2ab + b^2)$, $(a \pm b)^3 = a^3 \pm 3a^2b + 3ab^2 \pm b^3)$.

Example 1-2-1: If $a = mul.\, 3 + 1$, show that $a^2 = mul.\, 3 + 1$, and $a^3 = mul.\, 3 + 1$.

Solution

Since $a = mul.\, 3 + 1$, we may write, $a = 3k + 1$, where $k = 0,1,2,3,4, \ldots..$ Then,

$$a^2 = (3k + 1)^2 = 9k^2 + 6k + 1 = \underbrace{(3k^2 + 2k)}_{integer} \times 3 + 1 = mul.\, 3 + 1$$

Similarly,

$$a^3 = (3k + 1)^3 = (3k)^3 + 3 \times (3k)^2 \times 1 + 3 \times (3k) \times 1^2 + 1^3$$

$$a^3 = 27k^3 + 27k^2 + 9k + 1 = \underbrace{(9k^3 + 9k^2 + 3k)}_{integer} \times 3 + 1 = mul.\, 3 + 1$$

Example 1-2-2: If $a = mul.\, 5 + 1$, show that $3a^2 + 3a - 1 = mul.\, 5$.

Solution

Since $a = mul.\, 5 + 1$, we may write, $a = 5k + 1$, where $k = 0,1,2,3,4, \ldots.$

Then,

$$3a^2 + 3a - 1 = 3(5k + 1)^2 + 3(5k + 1) - 1$$

$$3a^2 + 3a - 1 = 3(25k^2 + 10k + 1) + 15k + 3 - 1$$

$$3a^2 + 3a - 1 = 75k^2 + 30k + 3 + 15k + 3 - 1 = 75k^2 + 45k + 5$$

$$3a^2 + 3a - 1 = \underbrace{(15k^2 + 9k + 1)}_{integer} \times 5 = mul.\,5$$

PROBLEMS

1-2-1) If $a = mul.\,5 + 2$, show that: $a + 3 = mul.\,5$, and $2a + 1 = mul.\,5$.

1-2-2) If $a = mul.\,5 + 3$, show that:

$$a^2 = mul.\,5 + 4, \quad a^3 = mul.\,5 + 2, \quad a^4 = mul.\,5 + 1$$

1-3) The equality of the algorithmic division

Let us start our analysis by considering a simple example. Let us take the two integers 43 and 8, and find **all the multiples of the smallest number 8, not exceeding 43**. These multiples are:

$$0 \times 8 = 0, 1 \times 8 = 8, 2 \times 8 = 16, 3 \times 8 = 24, 4 \times 8 = 32, 5 \times 8 = 40$$

The largest multiple of 8, not exceeding 43, is the number $40 = 5 \times 8$. This multiple differs from 43 by a number less than 8, since the next multiple $6 \times 8 = 48$ exceeds 43. The difference $43 - 40 = 3$ is the so called "**the remainder of the division of 43 by 8**". The equality $43 = 5 \times 8 + 3$ breaks down the number 43 into two parts, the first one being a multiple of 8, $(40 = 5 \times 8)$, and the second one, (remainder $= 3$), being smaller that 8.

The aforementioned analysis may be generalized, for any two positive integers, as it is shown in the following theorem.

Theorem 1-1: Given two integers a and b, with $b \neq 0$, there exists a unique pair of integers q and r such that:

$$a = qb + r, \quad with \ \ 0 \leq r < b \qquad\qquad (1-3-1)$$

Proof: a) Let us consider all the multiples of b, not exceeding a, and let us call qb the greatest of these multiples. This multiple qb will differ from a by a number less than b, i.e. $(a - qb) < b$. This is so since, the next to the qb multiple, which is $(q + 1)b$, will exceed a, i.e.

$$(q + 1)b > a \Rightarrow qb + b > a \Rightarrow b > a - qb, \quad or, \quad a - qb < b$$

So, if we call $r = a - qb$, it follows that

$$a = qb + r, \quad with \quad 0 \le r < b$$

So far we have shown the existence of the integers q (**quotient**) and r (**remainder**).

b) For the uniqueness: Assuming that there exists another pair Q (quotient) and R (remainder), we must have:

$$a = Qb + R, \quad with \quad 0 \le R < b \qquad (1 - 3 - 2)$$

From (1-3-1) and (1-3-2) it follows

$$qb + r = Qb + R \Rightarrow b(q - Q) = R - r \qquad (*)$$

Since $0 \le R < b$ and $0 \le r < b$, i.e., $-b < -r \le 0$, it follows (by addition)

$$-b < R - r < b \Longleftrightarrow |R - r| < |b| \qquad (**)$$

Taking the absolute values of both sides of equation (*) we find

$$|b(q - Q)| = |R - r| \Rightarrow |b||q - Q| = |R - r| \overset{(**)}{\Longrightarrow}$$

$$|b||q - Q| < |b| \Rightarrow |q - Q| < 1 \qquad (***)$$

Since q and Q are integers, their difference is another integer, and the only integer whose absolute value is less than one, (from eq. (***)), is the number zero, i.e., $q - Q = 0$, i.e. $q = Q$, and then, from (*), $r = R$. We have thus shown that the pair q and r is unique.

Equation (1-3-1) is known as the "**equality of the algorithmic division**", or, for brevity, the division of a by b. The operation of the division is denoted as $a \div b$.

The number a is called the **dividend**; the number b is the **divisor**, while q and r are the **quotient** and the **remainder** of the division, respectively.

In case the remainder $r = 0$, the division is called **perfect**. In this case, $(r = 0)$, $a = qb$, i.e. $a = mul. b$. If $r \neq 0$, the division is **not perfect**.

Comments: 1) The remainder r of the division $a \div b$ is the smallest positive integer, which when subtracted from a, makes it a multiple of b.

2) The remainder of the division $a \div b$ is not affected, if a is increased or decreased by a multiple of b.

This is so, since if kb is any multiple of b, then, from $a = qb + r$, it follows that

$$a + kb = qb + r + kb = (q + k)b + r$$

This equality shows that when $a + kb$ (a increased or decreased by a multiple of b) is divided by b, the remainder is still r, (not affected). The quotient of the division $(a + kb) \div b$, however, changes to $q + k$.

3) If the dividend is smaller than the divisor, $(a < b)$, then the quotient of the division $a \div b$ is $q = 0$ and the remainder is $r = a$, i.e.

$$a = 0 \times b + a \quad if \quad a < b$$

For example, if we consider $a = 5$ and $b = 9$, we may write, $5 = 0 \times 9 + 5$, (quotient $q = 0$ and remainder $r = 5 = a$).

4) Even and Odd numbers: If an integer number is divided by 2, the remainder of the division will be either 0 or 1, (since $0 \leq r < 2$, i.e. $r = 0$ or $r = 1$). Thus the set of integers is divided into two subsets. In the first subset belong all the numbers yielding remainder 0, when divided by 2, i.e. all the numbers which are divided exactly by 2. In the second subset belong all the numbers which, when divide by 2, yield remainder 1.

Even numbers: $\{0, 2, 4, 6, 8, 10, 12, \ldots \}$.

Odd numbers: $\{1, 3, 5, 7, 9, 11, 13, \ldots \}$.

Every even number is of the form $2k$, where k is an integer.

Every odd number is of the form $2k + 1$, where k is an integer.

1-4) Divisors of integers

We say that the integer d divides another integer a whenever $a = kd$, with k being some integer. In other words, d divides a if a is a multiple of d. In this case we write $d \, / \, a$. If d does not divide a, we write $d \nmid a$.

For example, the numbers 1,2,4,5,10,20 are all divisors of 20. We may write $1/20, 2/20, 4/20$, etc. On the other hand, $3 \nmid 20, 7 \nmid 20, 12 \nmid 20$, etc.

Properties:

1) For any integer a ($a \neq 0$) the following properties are obvious:

$$1/a, \quad a/a, \quad a/0$$

2) If an integer is a divisor of two other integers, then is a divisor of the sum and the difference of these two integers, i.e. if $d \, / \, a$ and $d \, / \, b$, then, $d \, / \, (a \pm b)$.

Indeed, $d \, / \, a$ means that $a = kd$, $d \, / \, b$ means $b = \lambda d$, with k and λ integers, and then $(a \pm b) = (k \pm \lambda)d$ which means that $d \, / \, (a \pm b)$.

3) If d is a divisor of one factor of a product of integers, then d is a divisor of the whole product.

Indeed, let us consider the product of integers abc. If $d \, / \, a$, then $a = kd$, k being an integer, and thus the product becomes, $abc = (kbc)d$, which shows that $d \, / \, (abc)$, since (kbc) is some integer.

Example 1-4-1: Show that: **a)** The product of two consecutive integers is an even number, **b)** The product of three consecutive integers is a multiple of 6.

Solution

a) If a is the first integer, then the next integer will be $(a + 1)$, and their product is $a(a + 1)$. The number a shall be either even or odd. If a is even, we may set $a = 2k$, with k being integer. Then

$$a(a + 1) = 2k(2k + 1) = 2\{k(2k + 1)\} \Longrightarrow 2 \, / \, a(a + 1)$$

If a is odd, we may set $a = 2k + 1$, and then

$$a(a + 1) = (2k + 1)(2k + 1 + 1) = (2k + 1)(2k + 2)$$
$$= 2\{(2k + 1)(k + 1)\} \Longrightarrow 2 \,/\, a(a + 1)$$

b) Let $a, a + 1, a + 2$ be three consecutive integers. Then, among these integers, one must be a multiple of 2 and one must be a multiple of 3. Then, the product $a(a + 1)(a + 2) = mul.\, 2 \times mul.\, 3 = mul.\, 6$, which means that $6 \,/\, a(a + 1)(a + 2)$.

Example 1-4-2: Show that the product of two odd numbers is another odd number.

Solution

Let $a = 2k + 1$ and $b = 2\lambda + 1$. Then, their product is

$$ab = (2k + 1)(2\lambda + 1) = 2\underbrace{(2k\lambda + k + \lambda)}_{integer\ n} + 1 = 2n + 1 = Odd$$

Example 1-4-3: The dividend of a division is 124, the quotient is 8 and the remainder is 4. Which is the divisor?

Solution

In our problem, $a = 124, q = 8$ and $r = 4$. The divisor b is to be determined. From the equality of the algorithmic division,

$$a = bq + r \Longrightarrow 124 = 8q + 4 \Longrightarrow 124 - 4 = 8q \Longrightarrow q = \frac{120}{8} = 15$$

Example 1-4-4: If $a \neq 0$ and $d \,/\, a$, show that $d \leq a$.

Solution

Since, by hypothesis, $d \,/\, a$, $a = kd$, with k some integer. The fact that $a \neq 0$, by hypothesis, implies that $k \neq 0$, which in turn implies that $k \geq 1$, and multiplying both sides by the positive integer d, yields, $kd \geq d$, or, $a \geq d$, (since $a = kd$).

Example 1-4-5: If $d \, / \, a$ and $d > a$, show that $a = 0$.

Solution

If a was not zero, i.e., if $a \neq 0$, then, according to example 1-4-4, $d \leq a$, which contradicts our hypothesis. Thus, necessarily, a must be zero.

Example 1-4-6: Show that the square of any odd integer is a multiple of 8 increased by 1.

Solution

If $a = 2k + 1$ is an odd integer, then, $a^2 = 4k^2 + 4k + 1 = 4k(k + 1) + 1$. By virtue of example 1-4-1, $k(k + 1) = mul. \, 2$, therefore, $4k(k + 1) = mul. \, 8$, and finally, $a^2 = mul. \, 8 + 1$, and this completes the proof.

Example 1-4-7: Find all the positive integers a, which when divided by another given positive integer b, yield a quotient equal to the remainder.

Solution

In general, we know that $a = bq + r$, where $0 \leq r < b$, (equality of the algorithmic division). Since now the remainder must be equal to the quotient, $(q = r)$, we obtain, $a = br + r = (b + 1)r$, with $r = 0, 1, 2, \dots, b - 1$. The sought for positive numbers are: $a = 0, b + 1, 2(b + 1), \dots, (b - 1)(b + 1)$.

Example 1-4-8: If $a = mul. \, 3 + 1$, show that: **a)** $a^2 = mul. \, 3 + 1$, **b)** $a^3 = mul. \, 3 + 1$, **c)** $a^n = mul. \, 3 + 1$, $n = 4, 5, 6, \dots$, **d)** Find the remainder of the divisions, $(208^3 \div 3)$ and $(1906^{51} \div 3)$.

Solution

a) Since $a = mul. \, 3 + 1$, we may set $a = 3k + 1$, with k some positive integer. Then, $a^2 = (3k + 1)^2 = 9k^2 + 6k + 1 = 3 \times (3k^2 + 2k) + 1 = mul. \, 3 + 1$, **b)** Since $a^2 = mul. \, 3 + 1$, we may set $a^2 = 3\lambda + 1$, with λ some positive integer, and thus

$$a^3 = a^2 \cdot a = (3\lambda + 1)(3k + 1) = 9k\lambda + 3k + 3\lambda + 1, \quad or$$

$$a^3 = 3 \times \underbrace{(3k\lambda + k + \lambda)}_{integer} + 1 = mul. \, 3 + 1$$

c) Since $a^3 = mul. 3 + 1$, we may set $a^3 = 3\rho + 1$, with ρ some positive integer, and thus

$$a^4 = a^3 \cdot a = (3\rho + 1)(3k + 1) = 3 \times \underbrace{(3\rho k + k + \rho)}_{integer} + 1 = mul. 3 + 1$$

Similarly, we show that $a^5, a^6, ...,$ etc, are all multiples of 3 increased by 1.

d) Since $208 = 3 \times 69 + 1 = mul. 3 + 1$, it follows, (from (b)), that $208^3 = mul. 3 + 1$, which shows that the remainder of the division of the number 208^3 by 3, is 1.

Similarly, since $1906 = 3 \times 635 + 1 = mul. 3 + 1$, $1906^{51} = mul. 3 + 1$, and the remainder of the division $1906^{51} \div 3$ is 1.

PROBLEMS

1-4-1) If the divisor of a division is 5, which are the possible remainders?

(Ans: 0, 1, 2, 3, 4).

1-4-2) Find all the numbers which when divided by 4 yield a quotient twice the remainder,   **(Ans:** 0,9,18,27).

1-4-3) If a, b, c, d, f are positive integers and if $a \,/\, f$ and $a \,/\, d$, show that $a \,/\, (bf + cd)$.

1-4-4) Show that the number 8 divides the product of two consecutive even integers.

1-4-5) Find the remainder of the following divisions: $286^{13} \div 3$, and, $148^{179} \div 3$.

(Ans: In both cases the remainder is 1, see example 1-4-8).

1-4-6) If the remainder of the division $a \div b$ is r and the quotient is q, show that the remainder of the division $(ka) \div (kb)$, with k some positive integer is kr, while the quotient remains unaffected (is q).

1-4-7) If two integers, when divided by a positive number k yield the same remainder, then these two numbers differ by a multiple of k, and conversely, if two integers differ by a multiple of k, then, both numbers when divided by k yield the same remainder.

1-5) Common divisors and the Greatest Common Divisor (G.C.D)

1) If a number d divides **both** integers a and b, then d is called **a common divisor** (c. d) of a and b. In general, two integers have **a finite number of common divisors**. Of all the common divisors, the greatest is called the greatest common divisor, (**G.C.D**) of the numbers a and b.

For example the divisors of the number 18 are: 1,2,3,6,9,18, while the divisors of 24 are: 1,2,3,4,6,8,12,24. The common divisors of the numbers 18 and 24 are: 1,2,3,6, and the greatest common divisor of 18 and 24 is the number 6, (the greatest of all the common divisors).

Note that 1 is a common divisor of any two integers.

The greatest common divisor of two integers a and b is denoted by (a, b). For example, as we have found, $(18,24) = 6$. Also, since the numbers 7 and 13 have no common divisors, except the number 1, we may write $(7,13) = 1$.

The following two theorems are helpful in determining the (G.C.D) of two integers.

Theorem 1-2: If the integer b divides the integer a, then the greatest common divisor of a and b is b. In symbols:

$$If \quad b/a \Longrightarrow (a, b) = b$$

Proof: Since b / b and b / a, (by assumption), b is a common divisor of a and b. Common divisor greater than b cannot exist, since it would not divide b. Therefore, b is the greatest common divisor of a and b, i.e. $(a, b) = b$.

Theorem 1-3: Let a and b be any two integers, with $a > b$. Then, the G.C.D. (a, b) of a and b is not affected if the largest integer a is replaced by the remainder r of the division of $a \div b$, i.e. $(a, b) = (b, r)$.

Proof: Let q and r be the quotient and the remainder, respectively, of the division $a \div b$, i.e.

$$a = qb + r \iff r = a - qb, \qquad 0 \leq r < b \qquad (*)$$

Every common divisor d of a and b divides the remainder r, i.e. d is also a divisor of r. Indeed, every common divisor d of a and b divides also qb, and therefore divides the difference $a - qb$, i.e. divides exactly the remainder r, (from (*)). Conversely, a common divisor of b and r, divides also qb, and therefore the sum $qb + r$, i.e. divides exactly a, (also from (*)).

We have thus shown that the pair of numbers a and b has exactly **the same divisors** with the pair of numbers b and r, and therefore, these two pairs of numbers have the same G.C.D, i.e. $(a, b) = (b, r)$.

This theorem suggests a method to determine the G.C.D of two given numbers, using successive divisions. Let us demonstrate the method by an example. Suppose we want to find the G.C.D of the numbers $a = 588$ and $b = 315$. By virtue of Theorem 1-3, we may replace the largest number 588 by the remainder of its division by 315. Performing the division $588 \div 315$, we find

$$588 = 1 \times 315 + 273$$

Thus, according to theorem 1-3, $(588,315) = (315,273)$.

Again, by virtue of the same theorem, we can replace 315 by the remainder of its division by 273. Performing the division $315 \div 273$ we find

$$315 = 1 \times 273 + 42$$

By theorem 1-3, $(315,273) = (273,42)$.

Performing the division $273 \div 42$, we find

$$273 = 6 \times 42 + 21$$

and thus, $(273,42) = (42,21)$. Now, since $21 \,/\, 42$, we have $(42,21) = 21$, by virtue of theorem 1-2.

In summary:

$$(588,315) = (315,273) = (273,42) = (42,21) = 21$$

This equality shows that the G.C.D of the numbers 588 and 315 is the number 21.

The method just described to find the G.C.D of two integers, is known as the "**Euclidean algorithm**".

In general, if a and b are two positive integers, with $a > b$, then we can find their G.C.D (a, b), by performing the following successive divisions:

$$a = bq + r, \quad where \ \ 0 \leq r < b$$

$$b = rq_1 + r_1, \quad where \ \ 0 \leq r_1 < r$$

$$r = r_1q_2 + r_2, \quad where \ \ 0 \leq r_2 < r_1$$

$$r_1 = r_2q_3 + r_3, \quad where \ \ 0 \leq r_3 < r_2$$

$$\cdots \quad \cdots \quad \cdots \quad \cdots \quad \cdots \quad \cdots \quad \cdots \quad \cdots \quad \cdots$$

$$r_{n-2} = r_{n-1}q_n + r_n, \quad where \ \ 0 \leq r_n < r_{n-1}$$

The successive divisions terminate when we reach at this division which renders remainder equal to zero. This will happen, sooner or later, since the remainders are integers which are continuously decreasing, ($r_n < r_{n-1} <$ $r_{n-2} < \cdots < r_1 < r$). Let us assume that the last equality renders zero remainder, i.e. $r_n = 0$. Then, the sought for $G.C.D$ of a and b is the number r_{n-1}, i.e. $(a, b) = r_{n-1}$. This is so, since, according to theorem 1-3, we have:

$$(a, b) = (b, r) = (r, r_1) = (r_1, r_2) = \cdots = (r_{n-2}, r_{n-1})$$

However, since $r_n = 0$, r_{n-1} divides exactly r_{n-2}, and according to theorem 1-2, the G.C.D of r_{n-1} and r_{n-2} is the number r_{n-1}, i.e. $(r_{n-2}, r_{n-1}) = r_{n-1}$, and therefore $(a, b) = r_{n-1}$ as well.

2) Properties of the G.C.D:

a) All the common divisors of two integers are also divisors of the G.C.D, and conversely, every divisor of the G.C.D is a common divisor of the two numbers.

Indeed, as it follows from the proof of theorem 1-3, all the common divisors of the pair (a, b) are also common divisors of the pair (r_{n-2}, r_{n-1}), and conversely. But all the common divisors of (r_{n-2}, r_{n-1}) divide $r_{n-1} = (a, b)$ which is the G.C.D of a and b.

b) If two positive integers a and b are multiplied by another positive integer λ, then their G.C.D is multiplied by λ, i.e.

If $(a, b) = c$, then $(\lambda a, \lambda b) = \lambda c$.

The proof follows easily if all the equalities in the Euclidean algorithm are multiplied by λ, (let the reader complete the proof).

c) If two positive integers a and b are both divided by one of their common divisors, let's say k, then their G.C.D is also divided by the same divisor k.

The proof also follows from the Euclidean algorithm, if each equality is divided by k.

3) G.C.D of more than two numbers.

Common divisors and G.C.D has any set of integers, $a, b, c, d, \ldots.$ For example, if we consider three integers a, b and c, then any integer which divides exactly the three numbers is called a common divisor of a, b and c. The greatest of all the common divisors, is the $G.C.D$ of a, b and c, and is denoted by (a, b, c). To find the G.C.D of a, b and c, we replace the first two by their G.C.D, let's say k, and then seek the G.C.D of k and c.

As an example, let us consider the numbers 12, 30 and 42.

The divisors of 12 are: 1, 2, 3, 4, 6, 12.

The divisors of 30 are: 1, 2, 3, 5, 6, 10, 15, 30.

The divisors of 42 are: 1, 2, 3, 6, 7, 14, 21, 42.

The common divisors of the three numbers 12, 30 and 42 are: 1, 2, 3 and 6, and the G.C.D is 6, i.e. $(12,30,42) = 6$.

Similarly we may find the G.C.D of four numbers, of five numbers, etc.

All the properties of the G.C.D for two numbers remain valid for the G.C.D of more than two numbers.

Example 1-5-1: Find the common divisors and then the G.C.D of 60 and 90.

Solution

Divisors of 60: 1, 2, 3, 4, 5, 6, 10, 12, 15, 20, 30, 60

Divisors of 90: 1, 2, 3, 5, 6, 9, 10, 15, 30, 45, 90

The common divisors of 60 and 90 are: 1, 2, 3, 5, 6, 10, 15, 30.

The G.C.D of 60 and 90, is $(60,90) = 30$.

Other method: (using the Euclidean algorithm).

Since $90 = 2 \times 60 + 30$, we have (by virtue of Theorem 1-3):

$$(90,60) = (60,30)$$

and since $30 \,/\, 60$, $(60,30) = 30$, (Theorem 1-2).

Example 1-5-2: Show that the G.C.D of the natural numbers a and b is also the G.C.D of the numbers $a + bc$ and $a + b(c + 1)$, where c is any natural number.

Solution

a) Let d be a common divisor of a and b. Then d divides bc and $b(c + 1)$, and therefore d divides the numbers $a + bc$ and $a + b(c + 1)$, i.e. d is a common divisor of the numbers $a + bc$ and $a + b(c + 1)$.

b) Conversely, let k be a common divisor of $a + bc$ and $a + b(c + 1)$. Then k will divide the difference of these two numbers, $a + b(c + 1) - \{a + bc\} = b$, i.e. $k \,/\, b$, and therefore divides bc as well. Now, since by assumption, k divides $a + bc$, it will divide also the difference $a + bc - bc = a$, i.e. $k \,/\, a$. So k is a common divisor of a and b.

We have thus shown that the two pairs of numbers, a and b, and $a + bc$ and $a + b(c + 1)$, have exactly the same common divisors, and therefore they will have the same G.C.D.

Example 1-5-3: If $k \in \mathbb{N}$, show that $\left(2k+1, \frac{k(k+1)}{2}\right) = 1$.

Solution

Let d be a common divisor of $2k+1$ and $k(k+1)/2$. Then d will divide the numbers $(2k+1)^2 = 4k^2 + 4k + 1$ and $8 \times k(k+1)/2 = 4k^2 + 4k$, and therefore, it will also divide their difference, i.e.

$$d/\{(4k^2 + 4k + 1) - (4k^2 + 4k)\} \Longrightarrow d/1 \Longrightarrow d = 1$$

So, the only common divisor of the two given numbers is the number 1, which is also their G.C.D.

PROBLEMS

1-5-1) Using the Euclidean algorithm find the G.C.D of the numbers 2093 and 462,   (**Ans:** $(2093,462) = 7$).

1-5-2) Find the G.C.D of the numbers 156, 212 and 316.

1-5-3) Find the largest integer k which when divides 3787 yields remainder 7, while when divides 1964 yields remainder 4,   (**Ans:** $k = 140$).

1-5-4) Find all the common divisors of the numbers 18900, 4410, 2100, and then find the G.C.D of these numbers.

1-5-5) Show that the G.C.D of a and b, divides exactly the G.C.D of a and bk, (a, b, k are natural integers, (use property a)).

1-6) Common multiples and the Least Common Multiple (L.C.M)

Let us consider the two numbers 4 and 7. Every other number which is divided exactly by 4 **and** by 7 is called "**a common multiple**" of 4 and 7. An obvious common multiple of 4 and 7, is their product $4 \times 7 = 28$. However, this is not the only common multiple of 4 and 7, since all the numbers, $2 \times (4 \times 7) = 56, 3 \times (4 \times 7) = 84, 4 \times (4 \times 7) = 112$, etc, are also multiples of 4 and 7. It is obvious that there is an infinite number of multiples of 4 and 7. The smallest of the common multiples is called "**the least common multiple**". In our example, the least common multiple of 4 and 7 is the number 28.

In general, if $a_1, a_2, a_3, \ldots, a_n$ are n natural numbers, then every natural number m which is divided exactly by a_1, **and** by a_2, **and** by a_3,..., **and** by a_n, is called **a common multiple** of $a_1, a_2, a_3, \ldots, a_n$. An obvious common multiple is the product of all the numbers, i.e. is the number $(a_1 \, a_2 a_3 \cdots a_n)$. But also, if k is any other natural number, then the number $k(a_1 \, a_2 a_3 \cdots a_n)$, is also a common multiple. We thus see that there is an infinite number of common multiples of the given numbers.

The smallest of the common multiples is called the "**least common multiple (L.C.M)**" and is denoted by $[a_1, a_2, a_3, \ldots, a_n]$.

For the L.C.M the following theorems are fundamental.

Theorem 1-4: All the common multiples of the natural numbers $a_1, a_2, a_3, \ldots, a_n$, are multiples of the L.C.M, (or, stated differently, the L.C.M is a divisor of all the common multiples).

Proof: Let M be the L.C.M of the given numbers and m be any other multiple of the same numbers. If we divide m by M we find:

$$m = qM + r, \quad or \quad \boldsymbol{r = m - qM}, \quad with \quad 0 \leq r < M$$

Since m and M are common multiples of the numbers $a_1, a_2, a_3, \ldots, a_n$, the number $m - qM$, i.e. the remainder $r = m - qM$, shall be another common multiple of the numbers, and since $0 \leq r < M$, (the L.C.M), r must necessarily be equal to zero, i.e. $\boldsymbol{r = 0}$, and therefore $m = qM$ which implies that $M \,/\, m$, i.e. the L.C.M is a divisor of m, and this completes the proof.

Theorem 1-5: If $\boldsymbol{a \,/\, b}$, then $[a, b] = b$.

The number b is a common multiple of a and b, since $a \,/\, b$, (by assumption), and also $b \,/\, b$. Any other common multiple of a and b would be greater than b. In other words, b is the smallest of all the common multiples of a and b, i.e. the L.C.M of a and b is the number b.

Example 1-6-1: Find the L.C.M of the numbers 10 and 50.

Solution

Since 50 is a multiple of 10 and at the same time a multiple of 50, the L.C.M is 50, (in symbols: $[10,50] = 50$).

Example 1-6-2: Find the L.C.M of the numbers 8 and 18.

Solution

We start with the largest number and form its successive multiples, until we reach a number which is divisible by the smaller number. This number is the sought for L.C.M. In our case, the multiples of 18, are: 18, 36, 54, 72, 90, ….The number 72 is divisible by 8, which means that $[8,18] = 72$.

Example 1-6-3: Find the L.C.M of the numbers 8, 12, 15 and 18.

Solution

We find the multiples of the largest number (18) and check which one of them is divisible by **all** the other numbers (8, 12, 15). This number shall be the L.C.M. In our case, we find that $[8,12,15,18] = 360$, (let the reader check it).

Example 1-6-4: Find the smallest positive integer k, which when divided by each number of the set $\{8,12,15,18\}$ yields remainder $r = 7$.

Solution

Since $k = mul.\, 8 + 7 = mul.\, 12 + 7 = mul.\, 15 + 7 = mul.\, 18 + 7$

it follows that

$$k - 7 = mul.\, 8 = mul.\, 12 = mul.\, 15 = mul.\, 18$$

This means that $k - 7$ is a common multiple of the numbers 8, 12,15, 18, and therefore, the smallest k must be the L.C.M. of these numbers, i.e.

$$k - 7 = [8,12,15,18] = 360 \Rightarrow k = 360 + 7 = 367$$

The sought for number is $k = 367$.

Comment: In Chapter 3, section 3-3, another method will be developed for the determination of the G.C.D and the L.C.M of a given set of numbers. This method is based on the factorization of numbers in terms of their prime divisors, (the fundamental theorem of Arithmetic).

PROBLEMS

1-6-1) Find the L.C.M of the numbers: **a)** 8 and 18, **b)** 12 and 15, **c)** 6, 9 and 15, **d)** 12, 18, 20 and 24,   (**Ans:** 72, 60, 90, 360).

1-6-2) Find the smallest positive integer k, which when divided by each number of the set $\{9,16,24\}$ yields remainder $r = 5$.

1-6-3) Find a positive integer $k < 380$, which when divided by 3 or by 4 or by 5 yields remainder 2, while when divided by 11 yields remainder zero.

(**Ans:** $k = 242$).

CHAPTER 2: PRIME NUMBERS (AN INTRODUCTION)

2-1) Numbers relatively prime (or Coprime)

1) Two natural numbers a and b are called **relatively prime**, (or coprime), if their G.C.D is equal to 1, i.e. if $(a, b) = 1$.

In other words, **two natural numbers are relatively prime if they do not have any common divisors, except the number 1.**

For example, the numbers 3 and 5 are relatively prime since $(3,5) = 1$, while 15 and 24 are not relatively prime since $(15,24) = 3$.

2) Many natural numbers $a_1, a_2, \ldots, a_n$ are called **relatively prime in pairs**, (or, pair wise relatively prime), if any two numbers are relatively prime, i.e. if $(a_k, a_\lambda) = 1$, **whenever $k \neq \lambda$.**

For instance, the numbers 3,7,10 and 19 are relatively prime in pairs, since

$$(3,7) = (3,10) = (3,19) = (7,10) = (7,19) = (10,19) = 1$$

The numbers 3,9,13 are **not** relatively prime in pairs, since $(3,9) = 3 \neq 1$.

3) Many natural numbers $a_1, a_2, \ldots, a_n$ are said to be **relatively prime in their entirety** if their G.C.D is 1, i.e. if $(a_1, a_2, \ldots, a_n) = 1$. For instance, the numbers 5, 10 and 9 are relatively prime in their entirety.

Theorem 2-1 (Euclid's theorem): If the number k divides the product ab and is relatively prime to a, then k will divide b, $(a, b, k$ are natural numbers).

In symbols: If $k \,/\, (ab)$ and $(a, k) = 1$, then $k \,/\, b$.

Proof: Since $(a, k) = 1$, it follows that $(ab, kb) = b$, (by virtue of property (b) in section 1-5). But, since $k \,/\, (ab)$, (by hypothesis), and $k \,/\, (kb)$, k which is a common divisor of ab and kb, must also divide the G.C.D. of these two numbers, (property (a) in section 1-5), i.e. k must divide b, (which is the G.C.D of ab and kb).

If, for instance, it is given that $3a = 10b$, the conclusion is that 3 must divide b and 10 must divide a, (since $(3,10) = 1$), (a, b natural numbers).

Theorem 2-2: If a number k is divided by several other numbers which are relatively primes in pairs, then the number k is divided by their product.

In symbols: If a_i/k, $i = 1, 2, 3, …, n$ and $(a_i, a_j) = 1$, whenever $i \neq j$, then $(a_1 a_2 a_3 …. a_n) / k$.

Proof: Let us, for simplicity, assume that k is divided by a and by b and by c, where a, b, c are pair wise prime numbers. Since a / k we may set $\boldsymbol{k = aq}$, where q is some natural number. By assumption, b divides k as well, i.e. b divides the number aq, (b / aq), and since a and b are relatively prime, in accordance to theorem 2-1, b must divide q, i.e. $\boldsymbol{q = mb}$, where m is some natural number, and thus $\boldsymbol{k = aq = abm}$. Again, by assumption, c / k, i.e. $c / (abm)$, and since c is relatively prime to a and b, it must divide m, i.e. $\boldsymbol{m = \lambda c}$, where λ is some natural number, and finally, $\boldsymbol{k = abm = (abc)\lambda}$. This equality shows that the product (abc) divides the number k, and this completes the proof.

For example if we know that a number k is divided by 4, by 5 and by 7, then the number k shall be divided by $(4 \times 5 \times 7) = 140$, (since 4, 5, 7 are relatively prime in pairs). If a number is divided by 3, 6 and 7, this does **not** necessarily imply that the number is divided by $(3 \times 6 \times 7) = 126$, since the numbers 3 and 6 are not relatively prime. For example, the number 42 is divided by 3, by 6 and by 7, but is not divided by $126 = 3 \times 6 \times 7$.

Theorem 2-3: If two natural numbers a and b are divided by their G.C.D, they become relatively prime.

Proof: Let d be the G.C.D of two natural numbers a and b. Then, by virtue of property (c) in section 1-5, the natural numbers a/d and b/d have G.C.D equal to $d/d = 1$, i.e. $(a/d, b/d) = 1$, and this shows that a/d and b/d are relatively primes .

For example, if we divide 12 and 15 by their G.C.D which is the number 3, we obtain 4 and 5, which are relatively primes. Also, if we divide 16 and 24 by their G.C.D = 8, we obtain 2 and 3, which are relatively primes.

Example 2-1-1: If k is a natural number, show: **a)** the numbers k and $k + 1$ are relatively prime, and **b)** the numbers $4k + 1$ and $5k + 1$ are relatively prime.

Solution

a) If d is a common divisor of k and $k + 1$, then, d should also be a divisor of the difference $(k + 1) - k = 1$, i.e. $d \, / \, 1$, and since the only divisor of 1 is 1, it follows that $d = 1$. Therefore, the only possible divisor of k and $k + 1$, (which at the same time is the G.C.D), is the number 1, i.e. $(k + 1, k) = 1$, and this shows that k and $k + 1$ are relatively prime numbers.

b) If d is a common divisor of $4k + 1$ and $5k + 1$, then, d should also be a divisor of the number: $5 \times (4k + 1) - 4 \times (5k + 1) = 1$, i.e. $d = 1$, and arguing as in (a), it follows that $4k + 1$ and $5k + 1$ are relatively primes.

Example 2-1-2: The product of three consecutive natural integers is always divisible by 6.

Solution

Let n be a natural number. We want to show that $n(n + 1)(n + 2) = mul. \, 6$.

We know that between two consecutive integers one is even, i.e. $n(n + 1)(n + 2) = mul. \, 2$, or, equivalently, $2 \, / \, n(n + 1)(n + 2)$.

We will now show that $3 \, / \, n(n + 1)(n + 2)$.

If we divide n by 3, we have: $n = 3q + r$, where $r = 0, 1, 2$ and q being some natural number (Euclidean division).

a) If $r = 0$, $n = 3q$, and $n(n + 1)(n + 2) = 3q(3q + 1)(3q + 2) = mul. \, 3$

b) If $r = 1$, $n = 3q + 1$, and $n(n + 1)(n + 2) = (3q + 1)(3q + 2)(3q + 3)$

$$n(n + 1)(n + 2) = 3(3q + 1)(3q + 2)(q + 1) = mul. \, 3$$

c) If $r = 2$, $n = 3q + 2$, and $n(n + 1)(n + 2) = (3q + 2)(3q + 3)(3q + 4)$

$$n(n + 1)(n + 2) = 3(3q + 2)(q + 1)(3q + 4) = mul. \, 3$$

So, we always have, $n(n + 1)(n + 2) = mul. 3$, or, $3 / n(n + 1)(n + 2)$

The number $n(n + 1)(n + 2)$ is divided by 2 **and** by 3, and since 2 and 3 are relatively prime numbers, $n(n + 1)(n + 2)$ shall be divided by their product $2 \times 3 = 6$, i.e. $6 / n(n + 1)(n + 2)$.

Comment: Similarly we may show that the product of five consecutive integers is divisible by 120, (see Pr. 2-1-1).

Example 2-1-3: If k is any natural number show that:
$$30 / k(k + 1)(2k + 1)(3k^2 + 3k - 1)$$

Solution

a) The given product is divisible by 2, (due to the term $k(k + 1)$).

b) The product is also divided by 3. To see why, we may write: $k = 3q + r$, $0 \leq r < 3$, (as we did in the previous example), and therefore, $k = 3q$, or $k = 3q + 1$, or $k = 3q + 2$. We may check easily that in all three cases, the product $k(k + 1)(2k + 1)(3k^2 + 3k - 1) = mul. 3$, or equivalently, $3 / k(k + 1)(2k + 1)(3k^2 + 3k - 1)$.

c) The product is also divisible by 5. To see why, we write: $k = 5\lambda + r$, where λ is some natural number and $r = 0, 1, 2, 3, 4$. If we consider the five cases, $k = 5\lambda$, $k = 5\lambda + 1$, $k = 5\lambda + 2$, $k = 5\lambda + 3$, $k = 5\lambda + 4$, we may check by direct calculations that the given product is always a multiple of 5, i.e. $5 / k(k + 1)(2k + 1)(3k^2 + 3k - 1)$, (for detailed calculations, see Pr. 2-1-2).

We have thus shown that the given product is divisible by 2, by 3 and by 5, and since 2, 3, and 5 are relatively prime in pairs, it is also divisible by the product $2 \times 3 \times 5 = 30$, (by virtue of Theorem 2-2).

PROBLEMS

2-1-1) Show that the product of five consecutive integers is divisible by 120.

2-1-2) Work out the details in Example 2-1-3.

2-1-3) If k is any natural number, show that: $3 / (k^3 - k)$.

2-1-4) For any natural number k, show that: $6 / k(k - 1)(2k - 1)$.

2-1-5) For any natural number k show that: $9/(k^3 + (k+1)^3 + (k+2)^3)$.

Hint: $k = 3q + r,\ where\ r \in \{0, 1, 2\}$.

2-1-6) If $3 \nmid k, (k \in \mathbb{N})$ and $a = 3^{2k} + 3^k + 1$, show that $13 / a$.

Hint: Since $3 \nmid k,\ k = 3q + 1,\ or,\ k = 3q + 2,\ (q \in \mathbb{N})$.

2-2) Prime numbers and composite numbers

1) A natural number p is said to be prime if it is not 1, (i.e. $p \neq 1$), and if its only divisors are 1 and p. In other words, prime numbers are the natural numbers greater than 1, which have only two divisors, the unit and themselves.

Every natural number $\neq 1$, which is not prime is called composite. In other words, a composite number, except 1 and itself, has other divisors as well.

For example the number 3 is a prime number, since its only divisors are 1 and 3. Similarly, 5 is a prime number, since its only divisors are 1 and 5. The same is true for the number 7. **The number 2 is the only even prime.**

On the contrary, the number 4 is composite, since it is divided by 2, i.e. the number 4 is divided by 1, by 4 and by 2. The number 15 is also composite, since it is divided by 3 and by 5, etc. Every even number, except 2, (which is prime), is a composite number, since it is divided by 2.

The first prime numbers, between 1 and 100, are the following:

2	3	5	7	11	13	17	19	23	29
31	37	41	43	47	53	59	61	67	71
73	79	83	89	97					

A first comment is that there is an irregularity in the way the prime numbers appear. If we know one prime number, there is no way to "predict" what the next prime will be.

Next theorem is of fundamental importance for our subsequent analysis.

Theorem 2-4: Every composite number has at least, one prime divisor.

Proof: A composite number, by definition, has divisors other than 1 and the number itself. The first divisor of a composite number, after the number 1, is called "**the second divisor**". In other words, the second divisor is **the smallest of all the divisors** of the composite number, which is greater than 1. (For example, the divisors of the composite number 15 are: 1, 3, 5, and 15. The second divisor is the number 3). If a is a composite number and d is its second divisor, **then d must be prime**. Since, if d were not prime, it would be composite, and as such, it would have a second divisor $d_1 < d$. But then, d_1 would also be a divisor of the composite number a as well, which implies that there exists a divisor of a, less than d (and $\neq 1$). However, this is not possible, since d was assumed to be the second divisor of a. Thus, d must necessarily be prime, and this completes the proof.

2) How to check whether a given natural number a is prime or composite.

We check by trials if the natural number a has some divisor p, which is prime and smaller than a. We start with the known prime numbers, 2, 3, 5, 7, 11, 13, 17, 19,..., until we find one of them which divides a. In this case, the number a is composite. If the trials are negative, until we reach a prime p whose square exceeds a, then the number a is prime, (for a justification, see theorem 3-4). In other words, **if a natural number a does not have a prime divisor $p \leq \sqrt{a}$, then a is a prime number**.

For example, let us check whether the number 157 is prime or not.

We find by successive trials that the number 157 is not divided by 2,3,5,7, 11 and 13, and since $13^2 = 169 > 157$, we conclude that 157 is a prime number.

If we consider the number 195, we find that this number is divided by 3, $(195 = 3 \times 65)$, and therefore 195 is a composite number.

One question that arises at this point is how many prime numbers are there? How many composite numbers are there? Answers to both questions will be given in chapter 3, section 3-4. For the time, we just mention that there are infinitely many prime numbers and also, infinitely many composite numbers.

2-3) Some important theorems on prime numbers

Theorem 2-5: Any two prime numbers are relatively prime, i.e. if p_1 and p_2 are any two prime numbers, then $(p_1, p_2) = 1$.

Proof: Obvious since two prime numbers do not share any common divisors, except the number 1.

Theorem 2-6: If a prime number p divides the product ab, (where a and b are natural numbers), then, p must divide at least one of the numbers a or b.

In symbols: If $p/ab \Rightarrow p/a \ or \ p \ / \ b$.

Proof: If $p \ / \ a$ the theorem is proved. If p does not divide a, then p and a will be relatively prime, and in accordance to Theorem 2-1, p must divide b.

Corollary 1: If a prime number p divides the product of natural numbers $abcd$ ..., then p will divide at least one of the numbers a, b, c, d, ...

Corollary 2: If $p \ / \ a^k$, (a and k natural numbers), then $p \ / \ a$.

Note that in theorem 2-6, the assumption that p is a prime, is essential. If p is not prime, the theorem may or may not be true. For example, 8 divides $24 = 4 \times 6$, but 8 does not divide neither one of the numbers 4 and 6. This is so, since 8 is not a prime number.

Theorem 2-7: a) If a and b are two natural numbers which are relatively prime, then any two powers of them will be relatively prime as well. In symbols:

If $(a, b) = 1$, then $(a^k, b^n) = 1$, (k, n are natural numbers).

b) If p_1, p_2 are two prime integers, and k and n are two natural numbers, then $(p_1{}^k, p_2{}^n) = 1$, i.e. any two powers of two prime integers are relatively prime.

Proof: a) Let us assume that a^k and b^n are not relatively prime, and let d be a common divisor of the numbers a^k and b^n. If d is a prime number $p \geq 2$, then, we must have, $p \ / \ a^k$ and $p \ / \ b^n$. In this case, corollary 2 implies that

p / a and p / b, i.e. $d = p \geq 2$ is a common divisor of a and b, which, however contradicts our hypothesis that $(a, b) = 1$. Thus, we conclude that a^k and b^n are relatively prime numbers, i.e. $(a^k, b^n) = 1$. If d is composite, then it must have a prime divisor, say q. Then, assuming that d is a common divisor of both a^k and b^n, the prime number q must also be a divisor of the same numbers, and reasoning as previously, we are led to a contradiction. Thus, in all cases, $(a^k, b^n) = 1$, provided that $(a, b) = 1$.

b) Since $(p_1, p_2) = 1$, (theorem 2-5), $(p_1{}^k, p_2{}^n) = 1$, according to part (a).

Theorem 2-8: If the natural number a is divided by the prime numbers $p_1, p_2, \ldots, p_n$, then a will be divided by the product $p_1 p_2 \cdots p_n$.

In symbols: If $p_1 / a, p_2 / a, \ldots, p_n / a$, then $(p_1 p_2 \cdots p_n) / a$.

Proof: The prime numbers $p_1, p_2, \ldots, p_n$ are relatively prime in pairs (theorem 2-5), and therefore, by virtue of theorem 2-2, the number a will be divided by the product $p_1 p_2 \cdots p_n$.

Theorem 2-9: Let , $p_1, p_2, \ldots, p_n$ be distinct prime numbers and $k_1, k_2, \ldots, k_n$ be natural numbers. Then, if each one of the numbers $p_1{}^{k_1}, p_2{}^{k_2}, \ldots, p_n{}^{k_n}$ divides the natural integer a, the product $p_1{}^{k_1} p_2{}^{k_2} \ldots p_n{}^{k_n}$ divides a as well.

In symbols: If $p_1{}^{k_1} / a, p_2{}^{k_2} / a, \ldots, p_n{}^{k_n} / a$, then $p_1{}^{k_1} p_2{}^{k_2} \ldots p_n{}^{k_n} / a$.

Proof: Since the numbers $p_1{}^{k_1}, p_2{}^{k_2}, \ldots, p_n{}^{k_n}$ are relatively prime in pairs, (theorem 2-7 (b)), and each one of these numbers divides a, their product must divide a as well, by virtue of theorem 2-2.

Theorem 2-10: Let $p_1, p_2, \ldots, p_n$ be distinct prime numbers. If a prime number p divides the product $p_1 p_2 p_3 \ldots p_n$, i.e. if $p / (p_1 p_2 p_3 \ldots p_n)$, then p must be equal to one of the $p_1, p_2, \ldots, p_n$.

Proof: According to corollary 1, since the given prime numbers are distinct, p must divide one of the factors $p_1, p_2, \ldots, p_n$, assume, for example, the first factor p_1, i.e. p / p_1, and since the prime p_1 has no other prime divisor except itself, it follows that $p = p_1$.

Corollary: If $p \, / \, p_1{}^n$, then $p = p_1$.

Example 2-3-1: The following numbers are prime or composite?

$323, 513, 2531, 3733$

Solution

We check, by trials, if the given number is divided by some prime number. If yes, the number is composite. If not, the number is prime. If the given number a is large, we try to find prime divisors p, such that $p < \sqrt{a}$, (see section 2-2).

Number 323: We find $\sqrt{323} \cong 17.97$, so we seek for prime divisors less than 17.97. The possible candidates, according to the table of prime numbers less than 100, (in section 2-2), are: 2, 3, 5, 7, 11, 13 and 17. We find that $17 \, / \, 323$, since $323 = 17 \times 19$, so 323 is composite.

Number 513: We find $\sqrt{513} \cong 22.64$, so we seek for prime divisors less than 22.64. The prime numbers less than 22.64, are: 2, 3, 5, 7, 13, 17, 19. We easily check that $19 \, / \, 513$, since $513 = 19 \times 27$, i.e. 513 is composite.

Number 2531: We find $\sqrt{2531} \cong 50.30$ and the sought for prime divisors of 2531, (if there exist any), must be less than 50.30. The possible candidates are: 2, 3, 5, 7, 13, 17, 19, 23, 29, 31, 37, 41, 43, 47. By direct trials we find that no one of these primes divides 2531, which means that the number 2532 is prime.

Number 3733: Working as previously, we find that 3733 is prime.

Example 2-3-2: If k is a natural number > 1, show that the number $k^4 + 4$ is composite.

Solution

$$k^4 + 4 = (k^2 + 2)^2 - 4k^2 = (k^2 + 2k + 2)(k^2 - 2k + 2) =$$

$$\{(k + 1)^2 + 1\}\{(k - 1)^2 + 1\}$$

For a given natural $k > 1$, each one of the numbers $\{(k + 1)^2 + 1\}$ and $\{(k - 1)^2 + 1\}$ is > 1, and this shows that $k^4 + 4$ is composite, (since it is expressed as a product of two other natural numbers > 1).

Example 2-3-3: Show that every prime number p greater than 3, (i.e. all the primes 5, 7, 11, 13, 17, 19, ...), is of the form: $mul.\,6 + 1$, or, $mul.\,6 - 1$.

Solution

Let p be a prime integer greater than 3. If we divide p by 3, we find:

$$p = 3q + r, \qquad r = 1, 2 \quad (with\ q\ some\ natural\ number) \qquad (*)$$

(Note that r **cannot be zero**, since then 3 would be a divisor of p, and p would not be prime, contrary to the hypothesis). So, the possible forms of p are: $p = 3q + 1$, or, $p = 3q + 2$, with q some integer.

a) If $p = 3q + 1$, q cannot be odd, since then p would be even, and therefore not prime. Indeed, for $q = 2k + 1$, with k some integer,

$$p = 3(2k + 1) + 1 = 6k + 4 = 2(3k + 2) = multiple\ of\ 2, (Not\ prime)$$

Therefore, necessarily, q must be even, say $q = 2\lambda$, (λ integer), and then,

$$p = 3(2\lambda) + 1 = 6\lambda + 1 = mul.\,6 + 1$$

b) If $p = 3q + 2$, q cannot be even, since then p would be even, and therefore not a prime. Indeed, for $q = 2k$, then
$$p = 3(2k) + 2 = 6k + 2 = 2(3k + 1) = mul.\,2, (Not\ prime)$$

Thus, q must necessarily be odd, i.e. $q = 2\lambda + 1$, and then

$$p = 3(2\lambda + 1) + 2 = 6\lambda + 5 = 6\lambda + 6 - 1 = 6(\lambda + 1) - 1 = mul.\,6 - 1$$

For example, $5 = 6 - 1$, $7 = 6 + 1$, $23 = 24 - 1$, $37 = 36 + 1$, etc.

Comment: The inverse is **not** necessarily true, i.e. $mul.\,6 \pm 1$ is **not** necessarily a prime number. For example, $36 = 6 \times 6 - 1 = 35 = 5 \times 7$, (composite number).

Example 2-3-4: Find all the prime numbers of the form $n^3 - 1$, $(n \in \mathbb{N})$.

Solution

Since $n^3 - 1 = (n - 1)(n^2 + n + 1)$, the number $n^3 - 1$ will be prime if and only if $(n - 1) = 1$, i.e. if $n = 2$. For all other values of n, $n^3 - 1$ would be expressed as the product of two other integers, each one being greater than 1, and therefore it would be a composite number.

For $n = 2$, we find $n^3 - 1 = 2^3 - 1 = 7$, and this is the only prime number of the form $n^3 - 1$.

Example 2-3-5: Are there any natural numbers k and λ such that the number $a = k^4 + 4\lambda^4$ be a prime number?

Solution

$$a = k^4 + 4\lambda^4 = (k^2)^2 + 4k^2\lambda^2 + (2\lambda^2)^2 - 4k^2\lambda^2 \Longrightarrow$$

$$a = (k^2 + 2\lambda^2)^2 - (2k\lambda)^2 = \{k^2 + 2\lambda^2 + 2k\lambda\}\{k^2 + 2\lambda^2 - 2k\lambda\} \quad (*)$$

If a is to be prime, then, the smallest factor of the product in (*) must be 1, i.e.

$$k^2 + 2\lambda^2 - 2k\lambda = 1, or, (k - \lambda)^2 + \lambda^2 = 1 \Longrightarrow \begin{pmatrix} k - \lambda = 0 \\ and \\ \lambda = 1 \end{pmatrix} \Longrightarrow \begin{pmatrix} k = \lambda \\ and \\ \lambda = 1 \end{pmatrix}$$

For $k = \lambda = 1$, the sought for prime number is $a = 1^4 + 4 \times 1^4 = 5$.

Example 2-3-6: If k is a natural number ≥ 3 and if one of the two numbers $2^k - 1$ and $2^k + 1$ is prime, then, show that the other one is composite.

Hint: a) If $k \in \mathbb{N}$, then: $x^n - y^n = mul.\,(x - y)$.

From the identity:

$$x^n - y^n = (x - y)(x^{n-1} + x^{n-2}y + x^{n-3}y^2 + \cdots + xy^{n-2} + y^{n-1})$$

it follows that $x^n - y^n = mul.\,(x - y)$.

b) If k is an **odd** natural number, then $x^n + y^n = mul.\,(x + y)$.

If n is odd, then $x^n + y^n = x^n - (-y)^n = mul.\,(x - (-y)) = mul.\,(x + y)$.

Solution

a) Let us assume that $2^k - 1$ is a prime number. Then the number k must be prime as well. Since, if k were a composite number ≥ 3, then, we could express k as the product of two other natural numbers, say n and m, i.e. we could write, $\boldsymbol{k = nm}$, with $n \geq 2, m \geq 2$. In this case,

$$2^k - 1 = 2^{nm} - 1 = (2^n)^m - 1 = mul.\,(2^n - 1)$$

But the number $mul.\,(2^n - 1)$, for $n \geq 2$, is composite, and this contradicts our hypothesis that $2^k - 1$ is prime. Thus, necessarily, k must be a prime number ≥ 3, and since any prime ≥ 3 is an odd number ≥ 3, the number $2^k + 1 = mul.\,(2 + 1) = mul.\,3$, which shows that $2^k + 1$ is composite.

b) Let us now assume that $2^k + 1$ is a prime number. Then, necessarily k must be an even number ≥ 3. Indeed, if k were an odd integer ≥ 3, then $2^k + 1 = mul.\,(2 + 1) = mul.\,3$, and then, $2^k + 1$ would be a composite number, contrary to our hypothesis. Thus, necessarily, k must be even, let's say $k = 2\lambda, = 2,3,4,5, \ldots$ Then,

$$2^k - 1 = 2^{2\lambda} - 1 = (2^2)^\lambda - 1 = 4^\lambda - 1 = mul.\,(4 - 1) = mul.\,3$$

This shows that $2^k - 1$ is composite.

PROBLEMS

2-3-1) Which of the following numbers are prime?

239, 499, 805, 827, 1123, 1381, 1512, 2689, 2610, 3413, 4012, 4793

(**Ans:** Prime numbers: 239, 499, 827, 1123, 1381, 2689, 3413, 4793, Composite numbers: 805, 1512, 2610, 4012).

2-3-2) Let a and b be two natural numbers, with $a > b$. If the number $a^2 - b^2$ is a prime number, show that $a^2 - b^2 = a + b$.

2-3-3) If p is a prime number > 2, find the natural number a, such that $(a^2 - p)$ to be the square of an integer. (**Ans:** $a = (p + 1)/2$).

Hint: From the requirement $a^2 - p = k^2$, it follows $(a - k)(a + k) = p$, and since, by hypothesis, p is a prime number, $a - k = 1$, i.e. $k = a - 1$, since, otherwise p would be composite, contrary to our hypothesis, etc.

2-3-4) If p, q and r are prime numbers each greater than 3, show that the number $p^2 + q^2 + r^2$ is composite.

Hint: As shown in ex. 2-3-3, each prime number > 3 is of the form $mul.\, 6 \pm 1$.

2-3-5) Show that there are no distinct prime numbers p_1, p_2, p_3, p_4 such that $p_1 p_2 = p_3 p_4$.

Hint: If $p_1 p_2 = p_3 p_4$, then p_1 should divide exactly the product $p_3 p_4$, i.e. $p_1 / p_3 p_4$, and by virtue of theorem 2-10, p_1 should be equal to one of p_3 or p_4, contrary to our hypothesis that the four prime numbers are distinct.

2-3-6) Show that the number $a = 2^k + 1$ is composite for k odd ≥ 3, and is prime for $k = 1$.

2-3-7) If k is a natural number ≥ 2, show that the number $a_k = 2^{2^k} - 6$ is a multiple of 10, and thus, is a composite number.

Hint: $a_{k+1} = 2^{2^{k+1}} - 6 = \left(2^{2^k}\right)^2 - 6 = (a_k + 6)^2 - 6 = a_k{}^2 + 12a_k + 30$. The number $a_2 = 2^{2^2} - 6 = 2^4 - 6 = 10$, and using the recursive formula, we find that a_3 is also a multiple of 10. By means of the same formula we may show that a_4 is a multiple of 10, and thus, proceeding step by step we show that $a_k = mul.\, 10$ for all $k \geq 2$.

2-3-8) Show that the only prime number of the form $p = n^3 + 1$ is the number 2.

Hint: $n^3 + 1 = (n + 1)(n^2 - n + 1) = (n + 1)(n(n - 1) + 1)$. For $n^3 + 1$ to be prime, the factor $n(n - 1) + 1$ must be 1, and this occurs when $n = 1$, etc.

CHAPTER 3: THE FUNDAMENTAL THEOREM OF ARITHMETIC

3-1) Introduction

We have learned, even in the elementary school, that every natural number, (whole number), can be expressed as a product of prime numbers. Of course, the examples we used in the elementary school, were limited to rather small number, for example:

$$8 = 2 \times 2 \times 2 = 2^3, \qquad 10 = 2 \times 5, \qquad 9 = 3 \times 3 = 3^2, \qquad 15 = 3 \times 5$$

$$21 = 3 \times 7, \qquad 36 = 2 \times 2 \times 3 \times 3 = 2^2 \times 3^2, \qquad 55 = 5 \times 11, \qquad etc$$

The representation of small whole numbers as a product of prime numbers seems to be obvious. However, what happens for sufficiently large whole numbers? Can we still represent sufficiently large numbers as a product of prime numbers? And if we can, is such a representation unique, or maybe there are two or more different such representations?

The answer is that any natural number, can indeed be represented as a product of prime numbers, and furthermore, this representation is unique, (there is one and only one representation of natural numbers as a product of prime numbers). This statement is actually the so called "**the fundamental theorem of arithmetic**", a rigorous proof of which will be given in the next section. It is called fundamental since it is the most important theorem of Arithmetic.

The first rigorous proof of the fundamental theorem of arithmetic, was given by the great German mathematician Carl Friedrich Gauss, in 1801, and was published in his book "**Disquisitions Arithmeticae**". This book had a revolutionary impact on the theory of numbers, and is still considered to be an all times classic.

3-2) The fundamental theorem of arithmetic

In this section we state and prove rigorously the fundamental theorem of arithmetic. This theorem states that every natural integer can be expressed as a product of prime numbers, and is one of the most important theorems with many applications.

Theorem 3-1: (The fundamental theorem of arithmetic)

Every natural number $n > 1$ can be represented as a product of prime numbers; moreover, such a representation is unique except for the order of the factors.

Proof: First we shall show that every natural number $n > 1$ can, indeed, be represented as a product of prime numbers. In the second part of the proof, we shall show that this representation is unique.

a) The representation of natural numbers as products of prime numbers is obvious for small numbers. For example, $4 = 2 \times 2 = 2^2$, $6 = 2 \times 3$, $8 = 2 \times 2 \times 2 = 2^3$, $9 = 3 \times 3 = 3^2$, $10 = 2 \times 5$, $12 = 2 \times 2 \times 3 = 2^2 \times 3$, $14 = 2 \times 7$, $15 = 3 \times 5$, etc. and proceeding similarly, step by step, we arrive at the conclusion that all the natural numbers, smaller than some natural number k, can be represented as product of prime numbers. **The choice of k is arbitrary**. (For example, in our previous example, we have shown that all the numbers smaller than 16 are represented as product of primes, and thus we may take $k = 16$).

Let us now consider a natural number $n \geq k$. If n is prime, the theorem is valid, (the product has only one factor). If n is not prime, it will be a composite number, and as we proved in theorem 2-4, it must have at least one prime divisor p_1, and in this case, $n = p_1 n_1$, with $n_1 < n$. If n_1 is prime, the theorem is proved. If n_1 is not prime, it will be composite, and as such it will have at least one prime divisor p_2, i.e. $n_1 = p_2 n_2$, and then, $n = p_1 p_2 n_2$, where $n_2 < n_1$. If n_2 is prime, the theorem is proved. If n_2 is not prime, it will be composite, and as such it will have at least one prime divisor p_3, i.e. $n_2 = p_3 n_3$, and then $n = p_1 p_2 p_3 n_3$, with $n_3 < n_2 < n_1 < n$. Working similarly, we reach a point where $n = p_1 p_2 p_3 \dots p_k n_k$, where n_k is either prime, and the theorem is proved, or, since the natural numbers $n_1, n_2, n_3, \dots, n_k$ keep decreasing, it will be $n_k < k$, in which case n_k is expressed as a product of prime numbers, i.e. $n_k = q_1 q_2 \dots q_\lambda$, ($q_1, q_2, \dots, q_\lambda$ are all prime numbers). So, finally, $n = p_1 p_2 p_3 \dots p_k q_1 q_2 \dots q_\lambda$, i.e. the natural number n has been expressed as a product of prime numbers, and this proves the theorem.

b) In part (a) we proved that every natural number n can be expressed as a product of prime numbers, and let us assume that: $n = p_1 p_2 p_3 \ldots p_m$, where $p_1, p_2, p_3, \ldots, p_m$ are all prime numbers. In this part we shall show that this representation is unique, except for the order of the factors.

Let us assume that n has two factorizations in prime numbers, say

$$n = p_1 p_2 p_3 \ldots p_m = q_1 q_2 q_3 \ldots q_s \qquad (*)$$

where, $p_1, p_2, \ldots, p_m, q_1, q_2, \ldots, q_s$ are all prime numbers. We will show that $m = s$, and that each p equals some q.

Indeed, since the first prime p_1, in (*), divides the left side, it will divide the right side as well, i.e. $p_1 \,/\, q_1 q_2 q_3 \ldots q_s$. By virtue of corollary 1, theorem 2-6, p_1 must divide at least one of the factors of the product $q_1 q_2 q_3 \ldots q_s$, let us say that it divides the first factor q_1, i.e. $p_1 \,/\, q_1$. However, since p_1 and q_1 are prime numbers, p_1 must be equal to q_1, i.e. $p_1 = q_1$, and now eq. (*) implies:

$$p_2 p_3 \ldots p_m = q_2 q_3 \ldots q_s \qquad (**)$$

Reasoning similarly, we find that $p_2 = q_2$, then, $p_3 = q_3$, etc.

So every factor in the left side of (*) appears in the right side of the same equation, and similarly every factor in the right side of (*) appears in the left side of the same equation. The two products in (*) are comprised of the same exactly factors, and are therefore identical.

Comments: 1) In the factorization of an integer, it is possible one prime factor to appear more than once, (for instance, $72 = 8 \times 9 = 2^3 \times 3^2$). If the distinct prime divisors of the natural number n are $p_1, p_2, \ldots, p_k$, and p_1 occurs a_1 times, p_2 occurs a_2 times,…, and p_k occurs a_k times, then we can write

$$n = p_1^{a_1} p_2^{a_2} p_3^{a_3} \cdots p_k^{a_k} \qquad (3-2-1)$$

This representation of n is called "**the standard form of the number**".

In eq. (3-2-1), the exponents $a_1, a_2, \ldots, a_k$ are natural numbers (1, 2, 3, …), i.e. $a_i \geq 1, i = 1, 2, 3, \ldots, k$.

The standard form, does not, of course, include all the prime numbers, but only, some of them. For instance, the standard form of the number 45 is $45 = 3^2 \times 5$. The prime number 2 is missing. Also, in the standard form of the number $539 = 7^2 \times 11$, the prime numbers $2, 3, 5$ are missing. But we may include these missing primes in the standard form of a number if we choose their exponents to be zero, i.e. we may write, for example, $45 = 2^0 \times 3^2 \times 5$, or, $539 = 2^0 \times 3^0 \times 5^0 \times 7^2 \times 11$, etc. In simple terms, we may include all the prime numbers in the standard form of a numbers, if we make the stipulation that a missing prime may appear with a zero exponent.

2) By virtue of the fundamental theorem of arithmetic, every natural number a can be expressed as $a = 2^k b$, where $k \geq 0$, (integer) and b an odd integer.

3) Even though every natural number can be expressed as a product of prime numbers, it is not always possible to find the prime divisors of a number, especially if the given number is sufficiently large. For instance, it has been proved that the numbers $2^{2^7} + 1$ and $2^{2^8} + 1$ are composite numbers, but their prime divisors are not known.

Example 3-2-1: Factorize the number 1260.

Solution

The procedure of factorization is represented schematically as below:

$$
\begin{array}{r|l}
1260 & 2 \\
630 & 2 \\
315 & 3 \\
105 & 3 \\
35 & 5 \\
7 & 7 \\
1 &
\end{array}
$$

The number 1260 is even, so its first prime divisor is 2. Dividing 1260 by 2, we find 630, which is also even, and its first prime divisor is again 2. Dividing 630 by 2 we find 315, which is odd. We check that 315 is divided exactly by 3, (the second prime divisor, after 2), the quotient being 105, which is again divisible by 3, giving a quotient 35. This number is divided by 5, yielding quotient 7, which when divided by 7, yields 1. Thus,

$$1620 = 2 \times 2 \times 3 \times 3 \times 5 \times 7 = 2^2 \times 3^2 \times 5 \times 7$$

Example 3-2-2: Factorize the number 51975.

Solution

Working similarly, as in the previous example we find:

51975	3
17325	3
5775	3
1925	5
385	5
77	7
11	11
1	

$$51975 = 3 \times 3 \times 3 \times 5 \times 5 \times 7 \times 11 = 3^3 \times 5^2 \times 7 \times 11$$

Example 3-2-3: Factorize the number 10! into its prime multipliers.
$$(10! = 1 \times 2 \times 3 \times 4 \times 5 \times 6 \times 7 \times 8 \times 9 \times 10)$$

Solution

The number 10!, (read 10 factorial), is defined to be the product of all the integers from 1 to 10 (included).

$$1 \times 2 \times 3 \times 4 \times 5 \times 6 \times 7 \times 8 \times 9 \times 10 =$$

$$1 \times 2 \times 3 \times (2^2) \times 5 \times (2 \times 3) \times 7 \times (2^3) \times (3^2) \times (2 \times 5), \quad i.e.$$

$$10! = 2^8 \times 3^4 \times 5^2 \times 7$$

Example 3-2-4: Find the smallest natural number k, such that the number $a = 120k$ to be the square of an integer. Which is this integer?

Solution

If we factorize the number 120 into its prime factors, we find:
$120 = 2^3 \times 3 \times 5$. Then, $a = 120k = 2^3 \times 3 \times 5 \times k$, and in order a to be a perfect square, the smallest value of k must be $k = 2 \times 3 \times 5 = 30$, since then:

$$a = 2^3 \times 3 \times 5 \times k = 2^3 \times 3 \times 5 \times 2 \times 3 \times 5 = 2^4 \times 3^2 \times 5^2$$
$$= (2^2 \times 3 \times 5)^2 = 60^2 = 3600$$

So, the sought for number $k = 30$, and the number $a = 60^2 = 3600$.

Example 3-2-5: If a prime number q divides the integer k^n, show that q^n divides k^n, (k and n are natural numbers).

Solution

In terms of its prime multipliers, let $k = p_1{}^{a_1} p_2{}^{a_2} \cdots p_\lambda{}^{a_\lambda}$. Then

$$k^n = (p_1{}^{a_1} p_2{}^{a_2} \cdots p_\lambda{}^{a_\lambda})^n = p_1{}^{na_1} p_2{}^{na_2} \cdots p_\lambda{}^{na_\lambda} \qquad (*)$$

Since, by assumption, $q \, / \, k^n$, i.e. $q \, / \, (p_1{}^{na_1} p_2{}^{na_2} \cdots p_\lambda{}^{na_\lambda})$ and q is prime, q must divide one of the factors, say the first factor $p_1{}^{na_1}$, i.e. $q \, / \, p_1{}^{na_1}$, and this, by virtue of the corollary of theorem 2-10, implies $q = p_1$. But then, $q^n = p_1{}^n$, and since $p_1{}^n \, / \, k^n$, as shown in equation (*), it follows that $q^n \, / \, k^n$.

PROBLEMS

3-2-1) Express the following numbers in their standard form:

90, 280, 2079, 2275, 1695694

(Ans: $90 = 2 \times 3^2 \times 5$, $280 = 2^3 \times 5 \times 7$, $2079 = 3^3 \times 7 \times 11$,

$2275 = 5^2 \times 7 \times 13$, $1695694 = 2 \times 7^2 \times 11^3 \times 13$).

3-2-2) Express the number 379456 in terms of its prime multipliers and then show that this number is a perfect square (of which number?). Repeat for the number 342225.

3-2-3) Find the smallest natural number k, such that the number $a = 945k$ to be the square of an integer. Which is this integer?   **(Ans:** $k = 105$, $a = 315^2$).

3-2-4) Express the number 13! in its canonical form, (see ex. 3-2-3).

3-2-5) Let a, b, λ be three natural numbers. If a^λ / b^λ, show that a / b.

Hint: Assume $a = p_1^{a_1} p_2^{a_2} \cdots p_n^{a_n}$ and $b = p_1^{b_1} p_2^{b_2} \cdots p_n^{b_n}$, where all the exponents are ≥ 0.

3-2-6) Find the positive and integer solutions x, y, z of the system:

$$\left\{ \begin{array}{c} 3^x + 3^y + 3^z = 981 \\ x < y < z \end{array} \right\}$$

(Ans: $x = 2, y = 5, z = 6$).

Hint: Express 981 in its standard form $3^2 \times 109$, and write the first equation as, $3^x(1 + 3^{y-x} + 3^{z-x}) = 3^2 \times 109$, from which $x = 2$ and $1 + 3^{y-x} + 3^{z-x} = 109$, i.e., $3^{y-2} + 3^{z-2} = 108 = 3^3 \times 2^2$, i.e. $3^{y-2}(1 + 3^{z-y}) = 3^3 \times 2^2$, from which $y - 2 = 3$, i.e. $y = 5$, etc.

3-3) G.C.D and L.C.M of numbers expressed in their standard form

1) The fundamental theorem of arithmetic provides an easy method to find the G.C.D and the L.C.M of several numbers, when they are expressed in their standard form, i.e. when expressed in terms of their prime divisors.

Let us start with an example. Find the G.C.D and the L.C.M of the numbers 378, 2100 and 2205. Let us first express the three numbers in their standard form, i.e. in terms of their prime divisors. We find:

$$378 = 2 \times 3^3 \times 7$$
$$2100 = 2^2 \times 3 \times 5^2 \times 7$$
$$2205 = 3^2 \times 5 \times 7^2$$

The G.C.D is **the largest** of the common divisors. The number 2 is not a common divisor, (it does not divide the number 2205). The number 3 is a common divisor of the three numbers. Note that the numbers 3^2 and 3^3 do not divide the number 2100, and therefore are not common divisors. The number 5 is not a common divisor since it does not divide the number 378. Finally, the number 7 is a common divisor of the three numbers, but 7^2 is not a common divisor, since it does not divide the first two numbers.

In summary, each one of the three numbers is divided by 3 and by 7, and since 3 and 7 are prime numbers, each number shall be divided by the product $3 \times 7 = 21$. This shows that 21 is a common divisor of the three numbers and, at the same time, is the largest divisor, in other words the number 21 is the G.C.D of the numbers: 378, 2100, 2205, i.e.

$$(378, 2100, 2205) = 3 \times 7 = 21$$

We may thus state the general rule: **The G.C.D of several integers, expressed as products of their prime divisors, is found by forming the product of the common divisors only, each raised to the lowest exponent.**

Reasoning similarly, we may state the following general rule for the L.C.M: **The L.C.M of several integers, expressed as products of their prime divisors, is found by forming the product of the common and the non-common divisors, each raised to the highest exponent.**

Thus, for the three numbers of our example, we have:

$$[378, 2100, 2205] = 2^2 \times 3^3 \times 5^2 \times 7^2 = 132300$$

2) A useful relation between the G.C.D and the L.C.M of two natural numbers a and b.

Let $a = p_1^{a_1} p_2^{a_2} \cdots p_n^{a_n}$ and $b = p_1^{b_1} p_2^{b_2} \cdots p_n^{b_n}$, where $a_k \geq 0$ and $b_k \geq 0$ for $k = 1, 2, \dots, n$, be two natural numbers expressed in their standard form. Then, the G.C.D of a and b is:

$$(a, b) = p_1^{min(a_1, b_1)} p_2^{min(a_2, b_2)} \cdots p_n^{min(a_n, b_n)} \qquad (*)$$

The L.C.M of a and b is:

$$[a, b] = p_1^{max(a_1, b_1)} p_2^{max(a_2, b_2)} \cdots p_n^{max(a_n, b_n)} \qquad (**)$$

Multiplying (*) and (*) together yields:

$$(a, b)[a, b]$$
$$= p_1^{min(a_1, b_1) + max(a_1, b_1)} p_2^{min(a_2, b_2) + max(a_2, b_2)} \cdots p_n^{min(a_n, b_n) + max(a_n, b_n)}$$
$$= p_1^{a_1 + b_1} p_2^{a_2 + b_2} \cdots p_n^{a_n + b_n} = ab$$

We are thus led to the following theorem:

Theorem 3-2: If (a, b) is the G.C.D of two natural numbers a and b, and $[a, b]$ is the L.C.M of the same two numbers, then $(a, b)[a, b] = ab$.

By means of this theorem, we may find the L.C.M if we know the G.C.D, since $[a, b] = ab/(a, b)$.

For example, if $a = 504 = 2^3 \times 3^2 \times 7$ and $b = 1470 = 2 \times 3 \times 5 \times 7^2$, then, the G.C.D is $(504, 1470) = 2 \times 3 \times 7 = 42$, and the L.C.M is $[504, 1470] = 2^3 \times 3^2 \times 5 \times 7^2 = 17640$. Note that:

$$(504, 1470)[504, 1470] = 42 \times 17640 = 740880 = 504 \times 1470$$

Example 3-3-1: Let $a, b \in \mathbb{N}$, and $d = (a, b) = n^2$, (the G.C.D of a and b), and $g = [a, b] = m^2$, (the L.C.M of a and b), $(n, m \in \mathbb{N})$. Show that each one of the numbers a and b is the square of a natural number.

Solution

Since $d = (a, b) = n^2$, we may set: $a = n^2 a_1$ and $b = n^2 b_1$, with $(a_1, b_1) = 1$. According to theorem 3-2,

$$g = [a, b] = m^2 = \frac{ab}{(a, b)} = \frac{(n^2 a_1)(n^2 b_1)}{n^2} = n^2 a_1 b_1$$

and since a_1 and b_1 are relatively primes, it follows that: $a_1 = k^2$ and $b_1 = \lambda^2$, $(k, \lambda \in \mathbb{N})$, and consequently, $a = n^2 a_1 = n^2 k^2 = (nk)^2$, and $b = n^2 b_1 = n^2 \lambda^2 = (n\lambda)^2$.

Example 3-3-2: Given that the L.C.M of 12 and of another number a is $[a, 12] = 180$, find all the possible values of a.

Solution

We have: $12 = 2^2 \cdot 3$ and $180 = 2^2 \cdot 3^2 \cdot 5$. Thus, a must be of the form, $a = 2^k \cdot 3^m \cdot 5^n$, where, $k = 0,1,2$, $m = 2$ and $n = 1$. The possible values of a are: $a = 2^0 \cdot 3^2 \cdot 5 = 45$, or $a = 2^1 \cdot 3^2 \cdot 5 = 90$, or $a = 2^2 \cdot 3^2 \cdot 5 = 180$.

Example 3-3-3: If $(a, b) = 1$, show that $[a, b] = ab$.

Solution

In its standard form: $a = p_1^{a_1} \cdot p_2^{a_2} \cdot \ldots \cdot p_k^{a_k}$ and $b = q_1^{b_1} \cdot q_2^{b_2} \cdot \ldots \cdot q_\lambda^{b_\lambda}$ where p_i $(i = 1,2, \ldots, k)$ and q_j $(j = 1,2, \ldots, \lambda)$ are prime numbers and all the exponents are non negative integers. Since the two numbers a and b are relatively prime, (by hypothesis $(a, b) = 1$), they do not share any common divisors, and therefore, their L.C.M is:

$$[a, b] = (p_1^{a_1} \cdot p_2^{a_2} \cdot \ldots \cdot p_k^{a_k}) \cdot \left(q_1^{b_1} \cdot q_2^{b_2} \cdot \ldots \cdot q_\lambda^{b_\lambda}\right) = ab$$

Comment: Another proof is by using theorem 3-2.

$$[a, b] = \frac{ab}{(a, b)} = \frac{ab}{1} = ab$$

3-4) Historical remarks

1) Euclid's Theorem

The discovery of the prime numbers and many of their elementary properties, are attributed to the ancient Greeks. Euclid began book VII of his "*Elements*" by defining a number as "**a multitude composed of units**", and a prime number as "**a number measured by a unit alone**", (**Πρώτος αριθμός εστίν ο μοναδι μόνη μετρούμενος**). Euclid was the first to prove that there are infinitely many prime numbers. His ingenious proof is presented below.

Theorem 3-3: (Euclid's theorem)

There are infinitely many prime numbers.

Proof: It suffices to show that for **every** prime integer p_n there exists another prime integer greater than p_n. Let us consider the first n prime numbers $p_1 = 2, p_2 = 3, \dots, p_n$, with $p_1 < p_2 < \cdots < p_n$, and form the number P as

$$P = p_1 p_2 p_3 \cdots p_n + 1 \qquad (*)$$

Note that P is not divisible by any one of the primes $p_1, p_2, p_3, \dots, p_n$, since P is a multiple of each one of them, increased by one. Therefore, there are two possibilities:

a) Either P is a prime number, greater than p_n, or

b) P is a composite number, and as such, according to the fundamental theorem of arithmetic, it will have a prime divisor greater than p_n.

In either case, we have shown that there exists a prime number greater than p_n, and this proves the theorem.

Corollary: If $p_1, p_2, p_3, \dots, p_n$ are the first n prime numbers, then, at least one prime number greater than p_n, lies between the numbers $(p_n + 1)$ and $(p_1 p_2 p_3 \cdots p_n + 1)$.

For example, let us consider the first three prime numbers $p_1 = 2, p_2 = 3$ and $p_3 = 5$. Then $(p_3 + 1) = 6$ and $(p_1 p_2 p_3 + 1) = 31$. Between the numbers 6 and 31 we have the prime numbers, 7, 11, 13, 17, 19, 23, 29, and 31.

It is obvious that also, **there are infinitely many composite numbers**, for instance, all the even numbers greater than 2, (i.e. 4, 6, 8, 10, ...), or, all the multiples of 3, greater than 3, (i.e. 6,9, 12, 15, ...), etc.

2) Sieve of Eratosthenes of Cyrene

Eratosthenes of Cyrene, a famous Greek astronomer and mathematician, is known mainly for his work on prime numbers and for measuring the diameter of the Earth.

The sieve of Eratosthenes is an algorithm, (a systematic procedure), for finding all the prime numbers, less that a given number. Let us demonstrate the method, by finding all the prime numbers less than 100. First we list all the numbers from 2 to 100, as shown in the following table. This table contains prime and composite numbers. The basic idea is to cross out, step by step, all the composite numbers, with the final table to contain only the prime numbers, less than 100.

Table 1: List of all integers from 2 to 100

	2	3	4	5	6	7	8	9	10
11	12	13	14	15	16	17	18	19	20
21	22	23	24	25	26	27	28	29	30
31	32	33	34	35	36	37	38	39	40
41	42	43	44	45	46	47	48	49	50
51	52	53	54	55	56	57	58	59	60
61	62	63	64	65	66	67	68	69	70
71	72	73	74	75	76	77	78	79	80
81	82	83	84	85	86	87	88	89	90
91	92	93	94	95	96	97	98	99	100

In this table, **the number 2 is the first prime integer**, (in fact, 2 is the only even prime, and all other primes are odd integers). **All the multiples of 2**, i.e.

the numbers 4, 6, 8, 10, 12, ... , 98, 100, are composite numbers. If we cross out the multiples of 2, we obtain the following table:

Table 2: All the multiples of 2, greater than 2, have been crossed out

	2	3		5		7		9	
11		13		15		17		19	
21		23		25		27		29	
31		33		35		37		39	
41		43		45		47		49	
51		53		55		57		59	
61		63		65		67		69	
71		73		75		77		79	
81		83		85		87		89	
91		93		95		97		99	

The number 3 is the second prime integer. **All the multiples of 3**, in table 2, are composite numbers. If we cross out all the multiples of 3, we obtain the following table:

Table 3: All the multiples of 3, greater than 3, have been crossed out

	2	3		5		7			
11		13				17		19	
		23		25				29	
31				35		37			
41		43				47		49	
		53		55				59	
61				65		67			
71		73				77		79	
		83		85				89	
91				95		97			

The number 5 is the third prime number. **All the multiples of 5**, in table 3 are composite numbers. If we cross out all the multiples of 5, we obtain the following table:

Table 4: All the multiples of 5, greater than 5, have been crossed out

	2	3		5		7			
11		13				17		19	
		23						29	
31						37			
41		43				47		49	
		53						59	
61						67			
71		73				77		79	
		83						89	
91						97			

The number 7 is the fourth prime number. All the multiples of 7, in table 4, are composite numbers. If we cross out all the multiples of 7, we obtain the following table:

Table 5: List of prime numbers less than 100

	2	3		5		7			
11		13				17		19	
		23						29	
31						37			
41		43				47			
		53						59	
61						67			
71		73						79	
		83						89	
						97			

Table 5 contains all the prime numbers less than 100. The method just described, is known as "**the sieve of Eratosthenes**". It can, of course, be used to find all the prime numbers less than any given natural number a.

Pertinent to the problem of finding the prime numbers less than a given number, is the problem of determining whether a given number a is prime or composite. As we have stated in section 2-2, **if a natural number a does not have a prime divisor $p \leq \sqrt{a}$, then a is a prime number**.

To justify this proposition, we consider first, the following theorem.

Theorem 3-4: Every composite number a is divided by at least one prime number p, such that $p^2 \leq a$.

Proof: Since a is composite, we may write $\boldsymbol{a = \beta \cdot \gamma}$, with $\beta > 1$ and $\gamma > 1$, and if we further assume that $\beta \leq \gamma$, then $\beta^2 \leq \beta \cdot \gamma$, i.e. $\beta^2 \leq a$. If we now consider a prime divisor p of β, then, since $p \leq \beta$, we shall have $p^2 \leq \beta^2$, and since $\beta^2 \leq a$, we finally conclude that $\boldsymbol{p^2 \leq a}$. However, since every divisor of β is a divisor of a as well, (since $a = \beta \cdot \gamma$), the prime p, which was assumed to be a prime divisor of β, is also a divisor of a, and this completes the proof.

As a consequence of theorem 3-4, we have the proposition, (attributed to Eratosthenes), that: "**a natural number a is prime, if it does not have prime divisor p with $p^2 \leq a$**", (or, equivalently, $\boldsymbol{p \leq \sqrt{a}}$).

3) Fermat numbers

The numbers of the form: $2^{2^n} + 1$, with $n = 0, 1, 2, 3, 4, ...$, are called "**the Fermat numbers**" and are symbolized by F_n . Back in 1640, Fermat conjectured that every number F_n is a prime number. However, despite his long term efforts, we could not prove his conjecture.

Fermat's conjecture is true for $n = 0, 1, 2, 3, 4$, since the numbers

$$F_0 = 3, \qquad F_1 = 5, \qquad F_2 = 17, \qquad F_3 = 257, \qquad F_4 = 65537$$

are indeed, prime numbers. However, as Euler proved in 1732, for $n = 5$, the number F_5 is a composite number, since $F_5 = 641 \times 6700417$. In 1880 it was proved that F_6 is also a composite number, and one of its divisors is the number 274177. Since then, it has been proved by several mathematicians that there are many composite Fermat numbers, for instance, all the numbers, from F_7 up to and including F_{16} are composite.

Eisenstein, one of the most beloved students of Gauss, conjectured that there are infinitely many Fermat prime numbers. However, today, it is believed that the set of the Fermat prime numbers is finite.

Fermat prime numbers appear in the following proposition, proved by Gauss, at the age of 18. "**A regular n-gon can be constructed with compass and straightedge if and only if $n = 2^k \cdot F_{\lambda_1} \cdot F_{\lambda_2} \cdots F_{\lambda_i}$, where k is a non-zero integer, ($k \geq 0$), and $F_{\lambda_1}, F_{\lambda_2}, \cdots, F_{\lambda_i}$ are distinct Fermat primes**".

For instance, for $k = 0, \lambda_1 = 0$, we have $n = 2^0 \cdot F_0 = 3$, (equilateral triangle), while for $k = 0, \lambda_1 = 1$, we have $n = 2^0 \cdot F_1 = 5$, (regular pentagon), etc.

We conclude this section by stating one important property of the Fermat numbers.

Theorem 3-5: Any two Fermat numbers are relatively primes.

Proof: Let $F_k = 2^{2^k} + 1$ and $F_{k+\lambda} = 2^{2^{k+\lambda}} + 1$, ($k \in \mathbb{N}$ and $\lambda \geq 1$), be any two Fermat numbers. Since

$$F_{k+\lambda} - 2 = 2^{2^{k+\lambda}} - 1 = \left(2^{2^k}\right)^{2^\lambda} - 1 \qquad (*)$$

if we set $a = 2^{2^k}$ and take into account that $a^{2^\lambda} - 1 = mul.(a + 1)$, (for the proof see Pr. 3-4-2), equation (*) implies:

$$F_{k+\lambda} - 2 = mul.\left(2^{2^k} + 1\right), \quad i.e. \quad F_{k+\lambda} - 2 = mul. F_k \qquad (**)$$

Now, if we assume that d is **a common divisor** of F_k and $F_{k+\lambda}$, then, from equation (**), it follows that d must also be a divisor of 2, i.e. $d = 1$ or $d = 2$, and since F_k is an odd number, the only allowed value of d is $\boldsymbol{d = 1}$, and this shows that F_k and $F_{k+\lambda}$ are relatively prime numbers, (they do not have common divisors, except $d = 1$).

Theorem 3-5 provides **another proof** that there are infinitely many primes. For, each one of the numbers $F_1, F_2, F_3, F_4, \ldots, F_n, \ldots$ has an odd prime divisor, (in accordance to the fundamental theorem of arithmetic), which is not a divisor of any other Fermat number, (since any two Fermat numbers do not share common divisors except $d = 1$). But, since there are infinitely many Fermat numbers, there must be infinitely many prime numbers as well.

4) The prime number theorem

a) According to Euclid's theorem there are infinitely many primes. **Question:** Is there a formula for the nth prime number p_n? ($p_1 = 2$). In other words, is there a formula, by means of which, we can find the prime number p_n for each value of the natural number n? The answer is negative, no such formula exists. Of course, there are empirical formulas yielding prime numbers for several values of n, but not for all n. For example, the polynomial $P(n) = n^2 - n + 41$ gives prime numbers for $n = 0, 1, 2, 3, \ldots, 38, 39, 40$, but for $n = 41, P(41) = 41^2 = 1681$ which is a composite number. This example is attributed to Euler. Similarly, the polynomial $Q(n) = n^2 - 79n + 1601$ gives prime numbers for $n = 0, 1, 2, 3, \ldots, 78, 79$, but for $n = 80, Q(80) = 1681 = 41^2$, which is a composite number.

b) As we have already stated, there is an irregularity in the distribution of the prime numbers. This becomes evident even for the first prime numbers less than 100, as shown in Table 5. Great mathematicians, Gauss and Legendre among others, made several attempts to unlock the secrets of the distribution of primes, unfortunately, with no success. They were, therefore forced, to attack the problem in a different way, by asking "**how many prime numbers do not exceed a given number x**". Let us denote by $\pi(x)$ the function which counts the number of primes $\leq x$, i.e.

$$\pi(x) = The \ number \ of \ prime \ numbers \ p, such \ that: 2 \leq p \leq x \qquad (*)$$

For example, $\pi(1) = 0, \pi(2) = 1, \pi(3) = 2, \pi(4) = 2, \pi(5) = 3, \pi(9) = 4$, $\pi(11) = 5, \pi(20) = 8, \pi(\sqrt{230}) = 6, \pi(100) = 25$, etc. Note that if p_n is the nth prime number, then $\pi(p_n) = n$, since there are exactly n prime numbers not exceeding p_n.

In 1896, the French mathematician **J. Hadamard** and the Belgian mathematician **de la Vallee Poussin**, independently and almost simultaneously proved that for sufficiently large values of x,

$$\pi(x) \cong \frac{x}{\ln x}, \quad (for \ sufficiently \ large \ values \ of \ x) \qquad (**)$$

The symbol $\cong$ is read: "**is approximately equal to**".

In more technical terms:

$$\lim_{x \to \infty} \frac{\pi(x)}{\left(\dfrac{x}{\ln x}\right)} = 1$$

Equation (**) is called "**the prime number theorem**". This remarkable theorem is one of the most important theorems in the theory of prime numbers.

c) Twin prime numbers

The pair of prime numbers: $(3,5), (5,7), (11,13), (17,19), (29,31)$, etc, (see Table 5), are called **twin primes**. It is not known whether there are infinitely many twin primes, i.e. pair of primes of the form p and $(p + 2)$ with p being a prime number, or, if their number is finite. This is still, an unsolved problem. In 1921, the Norwegian mathematician **Viggo Brun**, proved the inequality

$$T(x) < \frac{100x}{(\ln x)^2} \qquad (***)$$

where $T(x)$ is the number of twin prime numbers, not exceeding the number x.

5) Some unsolved problems concerning prime numbers

a) There are many unsolved problems related to the prime numbers. Perhaps, the most famous problem is **Goldbach's conjecture**, which is one of the oldest unsolved problems in mathematics. It states that, **every even integer > 2 is the sum of two prime numbers**. For small numbers this is easily verified, for instance, $4 = 2 + 2$, $6 = 3 + 3$, $12 = 5 + 7$, $16 = 13 + 3$, etc. However, even though the conjecture seems to be "reasonable and obvious", yet, there is no general proof for large even numbers. Of course, if we add two prime numbers > 2, the result will be an even number, since all primes > 2 are odd integers. However, is the reverse proposition true, i.e. is **every even number** greater than 2, expressible as the sum of two prime numbers? This is, in fact, the essence of Goldbach's conjecture.

b) Are there infinitely many twin primes of the form p and $p + 2$, with p being a prime number, or, their number is finite?

c) The set of Fermat numbers $F_k = 2^{2^k} + 1$ which are primes, is finite or infinite? Also, it is not known whether the set of the composite Fermat numbers is finite or infinite.

d) If n is an integer, are there infinitely many primes of the form $n^2 + 1$?

e) The Mersenne numbers are numbers of the form $M_p = 2^p - 1$, with p being a prime integer. These numbers were introduced by the French monk **Marin Mersenne**, who studied them in the early 17th century. In 1644, Mersenne conjectured that the only primes p not exceeding 257, for which M_p are prime numbers, are: $2, 3, 5, 7, 13, 17, 19, 31, 67, 127$ and 257. This conjecture turned out to be false, since, as it was proved, the numbers M_{67} and M_{257}, are composite numbers, while the numbers M_{61}, M_{89} and M_{107} are prime numbers. The following two problems, concerning Mersenne numbers, are still unsolved:

Are there infinitely many Mersenne prime numbers M_p, with p being some prime integer?

Are there infinitely many composite Mersenne numbers M_p, with p being some prime number?

Example 3-4-1: If $P_k = 2^k + 1$ is a prime number, (with k being a natural number > 1), show that $k = 2^n$.

Solution

Since $P_k = 2^k + 1$ is a prime number, k must be an even number, since with k odd, P_k would be divisible by 3, i.e. it would be a composite number, (see identity (b) in example 2-3-6). So, k must be of the form $\boldsymbol{k = 2^n \beta}$, with β being an odd number. But β must necessarily be 1, i.e. $\boldsymbol{\beta = 1}$, since if $\beta \neq 1$, then, $P_k = 2^k + 1 = 2^{2^n\beta} + 1 = \left(2^{2^n}\right)^{\beta} + 1 = mul.\,(2^{2^n} + 1)$, (since β is odd), and finally P_k would have a divisor of the form $2^{2^n} + 1$, and consequently it would be a composite number, contrary to our hypothesis. Thus, necessarily $\beta = 1$, and thus $k = 2^n$.

Example 3-4-2: If $M_k = a^k - 1$ is a prime number, $(a, k$ are natural numbers $> 1)$, then $a = 2$ and k is a prime number.

Solution

If $a > 2$, then $M_k = a^k - 1 = mul. (a - 1)$, and then, M_k would have as one of its divisors the number $(a - 1) \neq 1$, and M_k would be a composite number, contrary to our hypothesis. Thus, necessarily $\boldsymbol{a = 2}$. With $a = 2$, $M_k = 2^k - 1$. If k is a composite number, say $k = n \cdot m$, with at least one of the two numbers ≥ 2, say $n \geq 2$, then $M_k = (2^n)^m - 1 = mul. (2^n - 1) \neq 1$, and thus M_k would be composite, contrary to our hypothesis. Thus, necessarily, k must be a prime number, and this completes the proof.

PROBLEMS

3-4-1) Verify that the polynomial $P(x) = x^2 - x + 41$ gives prime numbers for $x = 0, 1, 2, 3, \dots, 39, 40$, but for $x = 41$ gives a composite number.

3-4-2) Show that $a^{2^\lambda} - 1 = mul. (a + 1)$.

Hint: $a^{2^\lambda} - 1 = (a^2)^{2^{\lambda-1}} - 1 = (a^2 - 1)(\dots) = (a + 1)(a - 1)(\dots)$

3-4-3) If $P_k = a^k + 1$ is a prime number, $(a, k$ natural numbers $> 1)$, show that a is even and k is of the form 2^n.

3-4-4) Use Table 5 to count the number of primes less than 100, and then, compare your answer with the number of primes predicted from the prime number theorem, (eq. (**)).

3-4-5) Using the prime number theorem, predict the number of primes less than 100,000,   **(Ans:** 8686 primes, approximately).

3-4-6) Is the number 683 prime or composite? Repeat for the number 697.

Hint: A natural number a is prime if it does **not** have a prime divisor $p \leq \sqrt{a}$. Applying this criterion, we find that 683 is prime, while 697 is composite.

3-5) On the rareness of the prime numbers

As we know, there are infinitely many prime numbers, (Euclid's theorem), but the way they are distributed is highly irregular. One observation we can make, though, is that prime numbers become rare, even by the time we get to 100, as seen from Table 5, and they become even rarer afterwards. It has been found that out of the first 100,000 numbers only 9,592 are prime numbers, (roughly 9,5 % of the first 100,000 numbers are prime numbers).

In general, studying large numbers we notice that the prime numbers become more and more rare, and in accordance to the following theorem, we can find two consecutive prime numbers that are further apart than any given number.

Theorem 3-6: For any given natural number n, there exist n consecutive composite numbers.

Proof: For any given natural number n, we consider the numbers

$$(n + 1)! + 2, (n + 1)! + 3, \ldots, (n + 1)! + n, (n + 1)! + n + 1$$

where $(n + 1)! = 1 \cdot 2 \cdot 3 \cdot \cdots \cdot n \cdot (n + 1)$, (read: $n + 1$ factorial). The number of these numbers is n, and they are all composite numbers, since for every $k = 2, 3, \ldots, n, n + 1$, the number k divides $(n + 1)! + k$. For instance, the number 2 divides $(n + 1)! + 2$, the number 3 divides $(n + 1)! + 3$, etc. Therefore, if consider two consecutive prime numbers p_λ and p_m, with $p_\lambda < (n + 1)! + 2$ and $p_m > (n + 1)! + n + 1$, (notice that in between $(n + 1)! + 2$ and $(n + 1)! + n + 1$ there are no prime numbers, since, as we proved all these numbers are composite), the distance between p_λ and p_m can be as large as we wish, provided that we choose n to be sufficiently large.

CHAPTER 4: THE NUMBER OF DIVISORS OF A NATURAL NUMBER - THE SUM OF THE DIVISORS OF A NATURAL NUMBER - PERFECT NUMBERS - AMICABLE NUMBERS

4-1) Introduction

In this chapter we shall develop a systematic method to find the number of divisors, and also, the sum of the divisors, of a given natural number. Let us start with a simple example. Suppose we want to find the divisors of the number 24. Obviously, the divisors of 24 are the numbers: 1, 2, 3, 4, 6, 8, 12 and 24. In total, the number 24 has 8 divisors. Having found the divisors of 24, we may now easily find their sum: 1+2+3+4+6+8+12+24=60.

Of course, in this case, the number 24 is a rather small number, and we can determine easily its divisors. However, for rather large numbers, the problem becomes quite complicated. For instance, imagine that we wish to find the divisors of the number 17,640, or an even greater number. This is not as easy as it was with the number 24.

In this chapter, we shall develop a general method, **based on the fundamental theorem of algebra**, in order to determine all the divisors and, also, the sum of all the divisors of any given natural number.

Let us illustrate the method starting again with the number 24. Expressing 24 in terms of its prime divisors, (by virtue of the fundamental theorem of arithmetic), we find that the standard form of 24 is: $\mathbf{24 = 2^3 \cdot 3}$.

The prime number 2 appears in the standard form of the number 24 in **(3+1=4)** ways, (including the number 1): $2^0, 2^1, 2^2, 2^3$, while the prime number 3 appears in **(1+1=2)** ways: $3^0, 3^1$. Any divisor of the number 24 results from the multiplication of **one** of the numbers $2^0, 2^1, 2^2, 2^3$ with **one** of the numbers $3^0, 3^1$. For example, the divisor $1 = 2^0 \cdot 3^0$, the divisor $3 = 2^0 \cdot 3^1$, the divisor $8 = 2^3 \cdot 3^0$, etc. We thus see that the number of all the divisors of the number 24 is $(3 + 1) \cdot (1 + 1) = 4 \cdot 2 = 8$, as found earlier.

To find the sum of all the divisors of the number 24, we first list all the divisors of 24, as shown below:

$$2^0 \cdot 3^0, 2^1 \cdot 3^0, 2^2 \cdot 3^0, 2^3 \cdot 3^0, 2^0 \cdot 3^1, 2^1 \cdot 3^1, 2^2 \cdot 3^1, 2^3 \cdot 3^1$$

The sum S of all the divisors of the number 24, is thus

$$S = 3^0 \cdot (2^0 + 2^1 + 2^2 + 2^3) + 3^1 \cdot (2^0 + 2^1 + 2^2 + 2^3)$$

$$S = (2^0 + 2^1 + 2^2 + 2^3) \cdot (3^0 + 3^1)$$

$$S = \frac{2^{3+1} - 1}{2 - 1} \cdot \frac{3^{1+1} - 1}{3 - 1} = \frac{2^4 - 1}{1} \cdot \frac{3^2 - 1}{2} = 15 \cdot 4 = 60 \qquad (*)$$

which is the same as the result we found earlier.

Note that, in deriving formula (*), we have used the identity

$$1 + x + x^2 + x^3 + \cdots + x^n = \frac{x^{n+1} - 1}{x - 1} \qquad (4-1-1)$$

The proof is easy. Let us call

$$S = 1 + x + x^2 + x^3 + \cdots + x^{n-1} + x^n \qquad (**)$$

Multiplying both sides of eq. (**) by x, results in the following,

$$xS = x + x^2 + x^3 + \cdots + x^n + x^{n+1} \qquad (***)$$

and if we subtract eq. (**) from eq. (***), we find:

$$(x - 1)S = x^{n+1} - 1, \;\; or, \;\; S = \frac{x^{n+1} - 1}{x - 1}$$

Of course, eq. (4-1-1) holds true provided that $x \neq 1$. But, if $x = 1$, the sum $S = n + 1$, as it follows readily from eq. (**).

In the next section, we will generalize the results found here, for an arbitrary natural number.

4-2) The number of divisors and the sum of the divisors of a natural number

a) The number of divisors: For any natural number a, let us call $\tau(a)$ the number of divisors of a. For instance, $\tau(2) = 2, \tau(6) = 4, \tau(15) = 4, \tau(24) = 8$, etc. If p is a prime integer, $\tau(p) = 2$, since the only divisors of p are 1 and p.

Theorem 4-1: If $a = p_1{}^{a_1} \cdot p_2{}^{a_2} \cdot \cdots \cdot p_k{}^{a_k}$ is the standard form of a natural number a, where $p_1 < p_2 < \cdots < p_k$ are prime numbers and $a_i \geq 0$, $i = 1, 2, \ldots, k$, are non-negative integers, then, the number of the divisors of a is given by the formula:

$$\tau(a) = (a_1 + 1)(a_2 + 1) \cdots (a_k + 1) \qquad (4-2-1)$$

Proof: Any divisor d of the number a, will be of the form

$$d = p_1{}^{b_1} \cdot p_2{}^{b_2} \cdot \cdots \cdot p_k{}^{b_k} \qquad (4-2-2)$$

where $0 \leq b_i \leq a_i$, for every $i = 1, 2, 3, \ldots, k$.

The prime number p_1 will appear in the expression of d in eq. (4-2-2) in $(a_1 + 1)$ ways, i.e. $1 = p_1{}^0, p_1{}^1, p_1{}^2, \cdots, p_1{}^{a_1}$. Similarly, the prime p_2 will appear in d in $(a_2 + 1)$ ways, ..., and the prime p_k will appear in d in $(a_k + 1)$ ways. Thus, all the prime divisors $p_1, p_2, p_3, \cdots, p_k$ will appear in all the factors of a in $(a_1 + 1)(a_2 + 1) \cdots (a_k + 1)$ ways.

For example, if $a = 12{,}600 = 2^3 \cdot 3^2 \cdot 5^2 \cdot 7$, then the number of divisors of a is: $\tau(12{,}600) = (3 + 1) \cdot (2 + 1) \cdot (2 + 1) \cdot (1 + 1) = 72$.

b) The sum of the divisors: For any natural number a, let us call $\sigma(a)$ the sum of all the divisors of a. For example, $\sigma(6) = 1 + 2 + 3 + 6 = 12$, $\sigma(10) = 1 + 2 + 5 + 10 = 18$, etc. If p is a prime integer, then $\sigma(p) = p + 1$, since the only divisors of p are the numbers 1 and p.

Theorem 4-2: If p is prime and n is any natural number, then

$$\sigma(p^n) = \frac{p^{n+1} - 1}{p - 1} \qquad (4-2-3)$$

Proof: The divisors of p^n are the numbers, $1 = p^0, p^1, p^2, \cdots, p^n$, and thus,

$$\sigma(p^n) = 1 + p^1 + p^2 + \cdots + p^n = \frac{p^{n+1} - 1}{p - 1}$$

according to formula (4-1-1).

Theorem 4-3: If two natural numbers n and m are relatively prime, i.e. if $(n, m) = 1$, then, $\sigma(nm) = \sigma(n)\sigma(m)$.

Proof: Since n and m are relatively primes, they do **not** share any common divisors. So, if we call $1 = n_1, n_2, n_3, \cdots, n_k = n$ the divisors of n, and $1 = m_1, m_2, m_3, \cdots, m_\lambda = m$ the divisors of m, since $(n, m) = 1$, all the divisors of the product nm will be of the form $\boldsymbol{n_i m_j}$, with $1 \le i \le k$ and $1 \le j \le \lambda$, and thus

$$\sigma(nm) = (n_1 m_1 + n_1 m_2 + \cdots n_1 m_\lambda) + (n_2 m_1 + n_2 m_2 + \cdots n_2 m_\lambda) + \cdots$$
$$+ (n_k m_1 + n_k m_2 + \cdots n_k m_\lambda) \Rightarrow$$

$$\sigma(nm) = n_1(m_1 + m_2 + \cdots + m_\lambda) + n_2(m_1 + m_2 + \cdots + m_\lambda)$$
$$+ \cdots n_k(m_1 + m_2 + \cdots + m_\lambda) \Rightarrow$$

$$\sigma(nm) = \underbrace{(n_1 + n_2 + \cdots + n_k)}_{\sigma(n)} \underbrace{(m_1 + m_2 + \cdots + m_\lambda)}_{\sigma(m)} = \sigma(n)(m)$$

and the proof is completed.

Comment: Similarly, if $(n, m) = 1$, then, $\tau(nm) = \tau(n)(m)$, (see Pr. 4-2-1).

Corollary: If n, m, λ are relatively primes in pairs, then, $\sigma(nm\lambda) = \sigma(n)\sigma(m)\sigma(\lambda)$. The proof follows easily, since the numbers (nm) and λ are relatively primes, i.e. they do **not** share any common divisors, and thus, according to theorem 4-3, $\sigma(nm\lambda) = \sigma((nm)\lambda) = \sigma(nm)\sigma(\lambda) = \sigma(n)\sigma(m)\sigma(\lambda)$. Extension to any number of factors is obvious, provided, of course, that **the factors are relatively primes in pairs**.

Theorem 4-4: If $a = p_1{}^{a_1} \cdot p_2{}^{a_2} \cdot \cdots \cdot p_k{}^{a_k}$ is the standard form of a natural number a, where $p_1 < p_2 < \cdots < p_k$ are prime numbers and $a_i \ge 0$, $i = 1, 2, \ldots, k$, are non-negative integers, then, the sum of the divisors of a is given by the formula:

$$\sigma(a) = \frac{p_1{}^{a_1+1} - 1}{p_1 - 1} \cdot \frac{p_2{}^{a_2+1} - 1}{p_2 - 1} \cdot \cdots \cdot \frac{p_k{}^{a_k+1} - 1}{p_k - 1} \qquad (4-2-4)$$

Proof: Follows readily from theorems 4-2 and 4-3.

For example, the sum of the divisors of the number $a = 36 = 2^2 \cdot 3^2$ is:

$$\sigma(36) = \frac{2^{2+1} - 1}{2 - 1} \cdot \frac{3^{2+1} - 1}{3 - 1} = \frac{7}{1} \cdot \frac{26}{2} = 91$$

Example 4-2-1: Find the number and the sum of the divisors, of the number $a = 630$.

Solution

$$a = 630 = 2 \cdot 3^2 \cdot 5 \cdot 7$$

$$\tau(630) = (1+1) \cdot (2+1) \cdot (1+1) \cdot (1+1) = 2 \cdot 3 \cdot 2 \cdot 2 = 24$$

$$\sigma(630) = \frac{2^{1+1} - 1}{2 - 1} \cdot \frac{3^{2+1} - 1}{3 - 1} \cdot \frac{5^{1+1} - 1}{5 - 1} \cdot \frac{7^{1+1} - 1}{7 - 1} = \frac{3}{1} \cdot \frac{26}{2} \cdot \frac{24}{4} \cdot \frac{48}{6} = 1872$$

Example 4-2-2: Find the natural numbers n of the form $n = 2^a \cdot 3^b$ which satisfy the equality: $\tau(n^2) = 3\tau(n)$.

Solution

With $n = 2^a \cdot 3^b$, $n^2 = 2^{2a} \cdot 3^{2b}$, $\tau(n) = (a+1)(b+1)$, $\tau(n^2) = (2a+1)(2b+1)$, and the given equality implies:

$(2a+1)(2b+1) = 3(a+1)(b+1)$, which is simplified to the following:

$$ab - a - b = 2, \ \ or, \ \ ab - a - b + 1 = 3, \ \ or, \ \ (a-1)(b-1) = 3 \qquad (*)$$

Since $(a-1)$ and $(b-1)$ are integers, the only possible solutions of the last equation in (*) are

$$a - 1 = 1 \ \textbf{and} \ b - 1 = 3, \ \textbf{or,} \ a - 1 = 3 \ \textbf{and} \ b - 1 = 1.$$

In the first case, $\boldsymbol{a = 2}$, $\boldsymbol{b = 4}$ and in the second case $\boldsymbol{a = 4}$, $\boldsymbol{b = 2}$, and therefore, the sought for numbers are: $n = 2^2 \cdot 3^4 = 324$, or, $n = 2^4 \cdot 3^2 = 144$.

Example 4-2-3: Find the natural number n of the form $n = p^a \cdot q^b$, with p and q prime integers, a and b natural numbers, such that $\tau(n) = 6$ and $\sigma(n) = 124$.

Solution

With $n = p^a \cdot q^b$, we have:

$$\tau(n) = (a+1)(b+1) = 6 \qquad (*)$$

$$\sigma(n) = \frac{p^{a+1} - 1}{p - 1} \cdot \frac{q^{b+1} - 1}{q - 1} = 124 \qquad (**)$$

Since a and b are natural numbers, eq. (*) implies that, $a + 1 = 2$ and $b + 1 = 3$, i.e. $a = 1$ and $b = 2$, or, $a + 1 = 3$ and $b + 1 = 2$, i.e. $a = 2$ and $b = 1$.

Case 1: $a = 1, b = 2$.

In this case, eq. (**) implies,

$$\frac{p^2 - 1}{p - 1} \cdot \frac{q^3 - 1}{q - 1} = 124 \Rightarrow (p + 1)(q^2 + q + 1) = 124 = 2^2 \cdot 31 \qquad (***)$$

We notice that the number $q^2 + q + 1$ is always an odd number. For, if $q = 2\lambda$ (even), then $q^2 + q + 1 = 4\lambda^2 + 2\lambda + 1 = 2(2\lambda^2 + \lambda) + 1$ is odd. If $q = 2\lambda + 1$ (odd), then $q^2 + q + 1 = 2(2\lambda^2 + 3\lambda + 1) + 1$, again an odd number. Thus, eq. (***) implies, $(p + 1) = 4$ and $(q^2 + q + 1) = 31$, i.e. $p = 3$ and $q^2 + q - 30 = 0$, or, $p = 3$ and $q = 5$.

In this case, $a = 1, b = 2, p = 3, q = 5$, the sought for number is $n = p^a \cdot q^b = 3^1 \cdot 5^2 = 3 \cdot 25 = 75$.

Case 2: $a = 2, b = 1$.

In this case eq. (**) implies,

$$\frac{p^3 - 1}{p - 1} \cdot \frac{q^2 - 1}{q - 1} = 124 \Rightarrow (p^2 + p + 1)(q + 1) = 124 = 2^2 \cdot 31$$

and working as in case 1, we find, $p = 5$ and $q = 3$.

In this case, $a = 2, b = 1, p = 5, q = 3$, we find $n = p^a \cdot q^b = 5^2 \cdot 3^1 = 75$.

Thus, the number $n = 75$ is the only natural number that satisfies the given conditions.

Example 4-2-4: Show that a natural number n is prime, if and only if, $\sigma(n) = n + 1$.

Solution

a) If n is a prime integer, its only divisors are 1 and n, and $\sigma(n) = n + 1$.

b) Conversely, if $\sigma(n) = n + 1$ we want to prove that n is a prime integer. If we assume that n is composite, then it must have **at least one** prime divisor $d \geq 2$, and in this case the divisors of n would be $1, d, n$, (at least), and then, $\sigma(n) \geq n + d + 1 > n + 1$, which contradicts our hypothesis. Thus, n must necessarily be a prime integer.

Example 4-2-5: If $n = 3^\lambda$ with λ integer ≥ 1, show that $2\sigma(n) = 3n - 1$.

Solution

By virtue of Theorem 4-2 we have:

$$\sigma(n) = \sigma\left(3^\lambda\right) = \frac{3^{\lambda+1} - 1}{3 - 1} = \frac{3 \cdot 3^\lambda - 1}{2} = \frac{3n - 1}{2} \Longrightarrow 2\sigma(n) = 3n - 1$$

Example 4-2-6: If Γ is the product of all the divisors $d_1 < d_2 < \cdots < d_n$ of the natural number a, show that $\Gamma^2 = a^n$.

Solution

Every natural number can be expressed as a product of its prime divisors.

a) Let us first assume that $a = p^k$, with p prime and $k \geq 1$. The divisors of a are the numbers: $1, p, p^2, \cdots, p^k$. The number of the divisors is $(k + 1)$, and their product is

$$\Gamma = 1 \cdot p \cdot p^2 \cdot \cdots \cdot p^k = p^{1+2+\cdots+k} = p^{\frac{k(k+1)}{2}} \Longrightarrow \Gamma^2 = p^{k(k+1)} = \left(p^k\right)^{k+1}$$

This shows that $\Gamma^2 = a^{k+1}$, (since $a = p^k$).

b) Let us now assume that $a = p_1{}^k p_2{}^\lambda$, with p_1, p_2 primes and k, λ integers ≥ 1. Every divisor d of the number a, will be of the form $p_1{}^i p_2{}^j$ with $0 \leq i \leq k$ and $0 \leq j \leq \lambda$. Let us, therefore, consider the following numbers:

$$1 = p_1^0, \qquad p_1^1, \qquad p_1^2, \qquad \cdots, p_1^k \qquad\qquad (*)$$

$$1 = p_2^0, \qquad p_2^1, \qquad p_2^2, \cdots, \qquad p_2^\lambda \qquad\qquad (**)$$

The divisors of a are the following:

$$1 \cdot 1, \qquad 1 \cdot p_2^1, \qquad 1 \cdot p_2^2, \qquad \cdots, \qquad 1 \cdot p_2^\lambda$$

$$p_1^1 \cdot 1, \qquad p_1^1 \cdot p_2^1, \qquad p_1^1 \cdot p_2^2, \qquad \cdots, \qquad p_1^1 \cdot p_2^\lambda$$

$$p_1^2 \cdot 1, \qquad p_1^2 \cdot p_2^1, \qquad p_1^2 \cdot p_2^2, \qquad \cdots, \qquad p_1^2 \cdot p_2^\lambda$$

$$\cdots \quad \cdots \quad \cdots \quad \cdots \quad \cdots \quad \cdots \quad \cdots \quad \cdots \quad \cdots \quad \cdots \quad \cdots \quad \cdots$$

$$p_1^k \cdot 1, \qquad p_1^k \cdot p_2^1, \qquad p_1^k \cdot p_2^2, \qquad \cdots, \qquad p_1^k \cdot p_2^\lambda$$

The total number of all the divisors is $(k + 1)(\lambda + 1)$.

The product Γ of all the divisors is:

$$\Gamma = \underbrace{\left\{1 \cdot \left(p_2^1 \cdot p_2^2 \cdot \cdots \cdot p_2^\lambda\right)\right\}}_{1^{st}\ row} \cdot \underbrace{\left\{p_1^{\lambda+1} \cdot \left(p_2^1 \cdot p_2^2 \cdot \cdots \cdot p_2^\lambda\right)\right\}}_{2^{nd}\ row} \cdot \cdots$$

$$\cdot \underbrace{\left\{p_1^{k(\lambda+1)} \cdot \left(p_2^1 \cdot p_2^2 \cdot \cdots \cdot p_2^\lambda\right)\right\}}_{(k+1)\ row} \Longrightarrow$$

$$\Gamma = \left\{p_2^1 \cdot p_2^2 \cdot \cdots \cdot p_2^\lambda\right\}^{k+1} \cdot \left\{p_1^{\lambda+1} \cdot p_1^{2(\lambda+1)} \cdot \cdots \cdot p_1^{k(\lambda+1)}\right\} \Longrightarrow$$

$$\Gamma = \left\{p_2^{1+2+\cdots+\lambda}\right\}^{k+1} \cdot \left\{p_1^{1+2+\cdots+k}\right\}^{\lambda+1} = p_1^{\frac{k(k+1)(\lambda+1)}{2}} \cdot p_2^{\frac{\lambda(\lambda+1)(k+1)}{2}} \Longrightarrow$$

$$\Gamma^2 = \left\{p_1^k p_2^\lambda\right\}^{(k+1)(\lambda+1)} = a^{(k+1)(\lambda+1)}$$

since $a = p_1^k p_2^\lambda$.

c) Similarly, proceeding step by step, we may prove the proposition for any natural number of the form $a = p_1^k \cdot p_2^\lambda \cdot p_3^m$, etc.

Example 4-2-7: Find the natural number a whose, the product of all its divisors, is $2^{30} \cdot 3^{40}$.

Solution

The number a will be of the form $a = 2^n \cdot 3^m$, (n, m natural numbers).

The total number of its divisors is $(n + 1)(m + 1)$, (Theorem 4-1). From the previous example, we must have:

$$(2^{30} \cdot 3^{40})^2 = (2^n \cdot 3^m)^{(n+1)(m+1)}, \quad or$$

$$2^{60} \cdot 3^{80} = 2^{n(n+1)(m+1)} \cdot 3^{m(n+1)(m+1)}$$

which implies that:

$$n(n + 1)(m + 1) = 60, \quad \textbf{and} \quad m(n + 1)(m + 1) = 80 \qquad (*)$$

Strictly speaking, the two equations in (*) form a system of two equations in the two unknown n and m. To solve the system, we divide the second by the first equation, and find: $m/n = 80/60 = 4/3$, and since 4 and 3 are relatively prime numbers, $\boldsymbol{m = 4k}$ and $\boldsymbol{n = 3k}$, with k integer. Then the first equation in (*) becomes:

$$3k(3k + 1)(4k + 1) = 60 \qquad (**)$$

We see that $k = 1$ satisfies the equation, and as a matter of fact is the only integer solution, (for the justification see Pr. 4-2-2). Thus, with $k = 1, \boldsymbol{n = 3}$ and $\boldsymbol{m = 4}$, and the sought for number is $a = 2^3 \cdot 3^4 = 648$.

4-2-8) Find two natural numbers a and b, such that: each one is divisible by 15 and each one having 10 divisors.

Solution

Let $a = 3^k \cdot 5^\lambda$ and $b = 3^m \cdot 3^n$, where $k, \lambda, m, n \in \mathbb{N}$. In this form, $15 / a$ and $15 / b$. Then:

$$\tau(a) = (k + 1)(\lambda + 1) = 10, \quad and \quad \tau(b) = (m + 1)(n + 1) = 10$$

From the first eq. we have, $k + 1 = 2$ and $\lambda + 1 = 5$, i.e. $\boldsymbol{k = 1, \lambda = 4}$, or, $k + 1 = 5$ and $\lambda + 1 = 2$, i.e. $\boldsymbol{k = 4, \lambda = 1}$, and similarly, from the second equation we find, $\boldsymbol{m = 1, n = 4}$, or, $\boldsymbol{m = 4, n = 1}$. Thus the two sought for numbers are: $a = 3^1 \cdot 5^4 = 1875$ and $b = 3^4 \cdot 5^1 = 405$.

PROBLEMS

4-2-1) If $(n, m) = 1$ show that: $\tau(nm) = \tau(n)(m)$.

4-2-2) In example 4-2-7, show that the only integer root of eq. (**) is $k = 1$, (as a matter of fact the other two roots are complex numbers).

4-2-3) Find the numbers: $\tau(1800)$ and $\sigma(1800)$.

(Ans: $\tau(1800) = 36$, $\sigma(1800) = 6045$).

4-2-4) Find the numbers: **a)** $\tau(100)$ and $\sigma(100)$, **b)** $\tau(1000)$ and $\sigma(1000)$, **c)** $\tau(1470)$ and $\sigma(1470)$, **d)** $\tau(9075)$ and $\sigma(9075)$.

4-2-5) If $a = p_1{}^k p_2{}^n p_3{}^m$ with p_1, p_2, p_3 prime numbers and k, n, m natural numbers, find the number of divisors and the sum of the divisors of a^2.

(Ans: $\tau(a^2) = (2k + 1)(2n + 1)(2m + 1)$

$$\sigma(a^2) = \frac{p_1{}^{2k+1} - 1}{p_1 - 1} \cdot \frac{p_2{}^{2n+1} - 1}{p_2 - 1} \cdot \frac{p_3{}^{2m+1} - 1}{p_3 - 1}$$

4-2-6) Repeat problem 4-2-5, for the number a^3.

4-2-7) Which natural number a of the form $a = 2^k \cdot 5^n \cdot 7^m$, $(k, n, m$ natural numbers), satisfy each one of the following equalities?

$$\tau(5a) = \tau(a) + 8, \quad \tau(7a) = \tau(a) + 12, \quad \tau(8a) = \tau(a) + 18$$

(Ans: $k = 3, n = 2, m = 1, a = 2^3 \cdot 5^2 \cdot 7^1 = 1400$).

4-2-8) Let $a = p_1{}^k p_2{}^n p_3{}^m$ be the standard form of a natural number a, $(p_1, p_2, p_3$ primes and k, n, m natural numbers). Show that the sum of the squares of all the divisors of a is given by the formula:

$$\frac{p_1{}^{2k+2} - 1}{p_1{}^2 - 1} \cdot \frac{p_2{}^{2n+2} - 1}{p_2{}^2 - 1} \cdot \frac{p_3{}^{2m+2} - 1}{p_3{}^2 - 1}$$

Hint: The **sum of the squares of all the divisors** of the number $a = p_1{}^k p_2{}^n p_3{}^m$, is equal to the product

$$\left(1 + p_1{}^2 + p_1{}^4 + \cdots p_1{}^{2k}\right)(1 + p_2{}^2 + p_2{}^4 + \cdots p_2{}^{2n})(1 + p_3{}^2 + p_3{}^4 + \cdots p_3{}^{2m})$$

Similarly, we may find **the sum of the cubes** of all the divisors of a, etc.

Comment: Notice that $\sigma(a^2)$, which is the sum of the divisors of the square a^2, (see Pr. 4-2-5), is different from the sum of the squares of the number a.

4-2-9) Find a natural number a of the from $a = 2^n \cdot 7^m$, given that the sum of the divisors of a is 56, while the sum of the squares of the divisors of a is 1050, **(Ans:** $a = 2^2 \cdot 7 = 28$).

Hint: Use the formula obtained in problem 4-2-8.

4-2-10) Find a natural number a of the form $a = 2^n \cdot 7^m$, given that the sum of the divisors of a is 24, while the sum of the cubes of the divisors of a is 3096, **(Ans:** $a = 2 \cdot 7 = 14$).

4-3) Perfect numbers

a) A proper divisor of a number is any divisor of the number other than the number itself. For example, all the divisors of the number 6 are the numbers $1, 2, 3, 6$. The proper divisors of 6 are the numbers $1, 2, 3$, (the number itself is excluded). Similarly, all the divisors of the number 24 are the numbers $1, 2, 3, 4, 6, 8, 12, 24$, while the proper divisors of 24 are: $1, 2, 3, 4, 6, 8, 12$, (the number 24 is excluded). **Let us denote by $\sigma_0(a)$ the sum of all the proper divisors of a.** For example, $\sigma_0(6) = 1 + 2 + 3 = 6$, while $\sigma_0(24) = 1 + 2 + 3 + 4 + 6 + 8 + 12 = 36$, etc. It is obvious that, for any natural number a, it is true that

$$\sigma(a) = \sigma_0(a) + a \qquad (4-3-1)$$

b) A natural number a is called perfect, if the number a is equal to the sum of all its proper divisors, i.e. if $\sigma_0(a) = a$, or, equivalently, by virtue of eq. (4-3-1), if

$$\sigma(a) = 2a \qquad (Perfect\ Number) \qquad (4-3-2)$$

For example, the number 6 is a perfect number, since $6 = 1 + 2 + 3$, $(\sigma_0(6) = 6)$. Another perfect number is the number 28, since the proper divisors of 28 are $1, 2, 4, 7, 14$ and $1 + 2 + 4 + 7 + 14 = 28$, $(\sigma_0(28) = 28)$.

Perfect numbers were studied by the ancient Greeks, including Euclid himself.

The next two perfect numbers are 496 and 8128. Today, we know that there are about 50 perfect numbers, all of which are even numbers. It is not known, yet, if odd perfect numbers exist.

Theorem 4-5: An even natural number a is perfect, if and only if is of the form $2^{p-1}(2^p - 1)$, with $2^p - 1$ prime, (p prime).

Proof

a) If $a = 2^{p-1}(2^p - 1)$, with $2^p - 1$ prime, then, from theorem 4-4,

$$\sigma(a) = \frac{2^p - 1}{2 - 1} \cdot \frac{(2^p - 1)^2 - 1}{(2^p - 1) - 1} = (2^p - 1) \cdot \{(2^p - 1) + 1\} = 2^p(2^p - 1)$$

i.e. $\sigma(a) = 2a$, meaning that a is a perfect number.

b) Now we shall prove the converse proposition. If a is an even perfect number, then a must necessarily be of the form: $2^{p-1}(2^p - 1)$, with $2^p - 1$ prime, (p prime).

Since a is even, we may write: $\boldsymbol{a = 2^{n-1}\beta}$, with β odd and $n \geq 2$, (see comment 2, in theorem 3-1). Then, since 2^{n-1} and β are relatively prime numbers, (2^{n-1} is even, while β is odd), theorem 4-3 implies:

$$\sigma(a) = \sigma(2^{n-1})\sigma(\beta) = \frac{2^n - 1}{2 - 1}\sigma(\beta) = (2^n - 1)\sigma(\beta) \qquad (*)$$

and since a was assumed to be a perfect number, $\sigma(a) = 2a = 2^n\beta$, i.e.

$$2^n\beta = (2^n - 1)\sigma(\beta) \qquad (**)$$

From eq. (**) it follows that $\beta = mul.(2^n - 1)$, since $(2^n - 1) / (2^n\beta)$ and is relatively prime to 2^n, and also that

$$\sigma(\beta) = \frac{2^n\beta}{2^n - 1} = \beta + \underbrace{\frac{\beta}{2^n - 1}}_{\gamma} = \beta + \gamma, \; where \; \gamma = \frac{\beta}{2^n - 1} \qquad (***)$$

Notice that γ is a proper divisor of β, (i.e. $\gamma < \beta$), since $n \geq 2$. Equation (***) implies, $\boldsymbol{\gamma = 1}$ **and** $\boldsymbol{\beta}$ **prime**, since the sum of the divisors of β, (i.e. $\sigma(\beta)$), is equal to the sum of β and its proper divisor γ. Consequently, since

$\gamma = 1, \beta = 2^n - 1$, (from eq. (***)), and thus $a = 2^{n-1}(2^n - 1)$, with $\beta = 2^n - 1$ **prime**. However, when $2^n - 1$ is prime, then, necessarily, n must be prime, say $n = p$ prime, (see ex. 3-4-2), and thus, $a = 2^{p-1}(2^p - 1)$, with $2^p - 1$ prime, (p prime).

Comments: 1) Part (a) of the proof is attributed to Euclid, and in his honor, the numbers of the form $2^{p-1}(2^p - 1)$, with $2^p - 1$ primes, (p prime), are called "**Euclid's numbers**", while part (b) of the proof, is attributed to Euler.

2) Recall that numbers of the form $2^p - 1$, with p prime, are the "**Mersenne numbers**", mentioned in section 3-4 (5).

Example 4-3-1: Verify that 496 is a perfect number.

Solution

The proper divisors of 496 are: 1, 2, 4, 8, 16, 31, 62, 124 and 248. The sum of all the proper divisors is: $1 + 2 + 4 + 8 + 16 + 31 + 62 + 124 + 248 = 496$, and this shows that 496 is a perfect number.

Example 4-3-2: Show that the sum of the reciprocals of all the divisors of an even perfect number is 2.

Solution

Let a be an even perfect number. According to theorem 4-5, a will be of the form: $a = 2^{p-1}(2^p - 1)$, with $2^p - 1$ prime, (p prime).

The divisors of a are:

$$2^0 = 1, 2, 2^2, \cdots, 2^{p-1}, 2^p - 1, 2 \cdot (2^p - 1), \cdots, 2^{p-1} \cdot (2^p - 1)$$

If we call S the sum of the reciprocals of all the divisors, we have:

$$S = 1 + \frac{1}{2} + \frac{1}{2^2} + \cdots + \frac{1}{2^{p-1}} + \frac{1}{2^p - 1} + \frac{1}{2 \cdot (2^p - 1)} + \cdots + \frac{1}{2^{p-1} \cdot (2^p - 1)}$$

$$S = \left(1 + \frac{1}{2} + \frac{1}{2^2} + \cdots + \frac{1}{2^{p-1}}\right) + \frac{1}{2^p - 1}\left(1 + \frac{1}{2} + \frac{1}{2^2} + \cdots + \frac{1}{2^{p-1}}\right)$$

$$S = \left(1 + \frac{1}{2} + \frac{1}{2^2} + \cdots + \frac{1}{2^{p-1}}\right)\left(1 + \frac{1}{2^p - 1}\right)$$

$$S = \left(1 + \frac{1}{2} + \frac{1}{2^2} + \cdots + \frac{1}{2^{p-1}}\right) \cdot \frac{2^p}{2^p - 1} \qquad (*)$$

The first factor in eq. (*) is:

$$1 + \frac{1}{2} + \frac{1}{2^2} + \cdots + \frac{1}{2^{p-1}} = \frac{1 - \left(\frac{1}{2}\right)^p}{1 - \frac{1}{2}} = \frac{2^p - 1}{2^{p-1}} \qquad (**)$$

By virtue of eq. (**), eq. (*) yields:

$$S = \frac{2^p - 1}{2^{p-1}} \cdot \frac{2^p}{2^p - 1} = 2$$

PROBLEMS

4-3-1) Find the natural number k, so that the number $a = 2^4 \cdot k$ to be a perfect number, (**Ans:** $k = 2^5 - 1 = 31$, $a = 496$).

4-3-2) Given that $a = 2^n \cdot 3^m$ is a perfect number, find the natural numbers n and m, (**Ans:** $n = 1, m = 1$, $a = 6$).

4-4) Amicable numbers

Two natural numbers a and b are called amicable, when each number is equal to the sum of the proper divisors of the other, (that is, the sum of all the divisors except the number itself). Amicable numbers are also called "**friendly numbers**". The smallest pair of amicable numbers is, 220 and 284.

The proper divisors of 220 are: $1, 2, 4, 5, 10, 11, 20, 22, 44, 55, 110$ and their sum is $\sigma_0(220) = 284$, (recall that $\sigma_0(a)$ denotes the sum of the proper divisors of a).

The proper divisors of 284 are: 1, 2, 4, 71, 142 and their sum is $\sigma_0(284) = 220$.

We have thus verified that 220 and 284 are amicable numbers.

The first five pairs of amicable numbers are:

$(220,284), (1184,1210), (2620,2924), (5020,5564), (6232,6368)$

There are about 390 known pairs of amicable numbers, but, still, it is not known if there are infinitely many pairs of them, or if their number is finite.

Amicable numbers were known to Pythagoreans (500 B.C), who credited them with many mystical properties, and it was believed that two persons, wearing amulets bearing the numbers 220 and 284, would be united with long and strong friendship. This mystical property attributed to the amicable numbers, justify the term "friendly numbers".

CHAPTER 5: DIVISIBILITY IN THE SET $\mathbb{Z}$ - THEOREMS ON THE REMAINDERS – CONGRUENCES

5-1) Introduction

In the preceding chapters we have dealt with the non-positive integers, i.e. with the numbers: $0, 1, 2, 3, 4, 5, \ldots \ldots$ Recall that $\mathbb{N} = \{1, 2, 3, 4, 5, \ldots\}$ is the set of **natural numbers or positive integers**, while $\mathbb{N}_0 = \{0, 1, 2, 3, 4, 5, \ldots\}$ is the set of natural numbers including 0, i.e. is the set of all the non-negative integers. The set $\mathbb{N}_0$ is also called "**the integers of arithmetic**".

The numbers: $\ldots, -5, -4, -3, -2, -1, 0, 1, 2, 3, 4, 5, \ldots$, are called "**the integers of algebra**", which include the positive integers, the negative integers and the number zero. The set of the integers of algebra is denoted by $\mathbb{Z}$, i.e. $\mathbb{Z} = \{\ldots, -5, -4, -3, -2, -1, 0, 1, 2, 3, 4, 5, \ldots\}$.

All the properties of the positive integers may be easily extended, with minor modifications, to the integers of algebra. Thus:

a) Divisibility: The definitions are identical to the definitions given in section 1-4, with the exception that, here, we may have negative divisors. For example, the divisors of 6 are the numbers: $1, -1, 2, -2, 3, -3, 6, -6$. The same numbers are also the divisors of the number -6.

We say that the integer d divides another integer a whenever $\boldsymbol{a = kd}$, with k being some integer, where, in this case, $a, k, d \in \mathbb{Z}$. For example $-3 / -6$ since $(-6) = 2 \cdot (-3)$. Also, $2 / (-8)$ since $(-8) = (-4) \cdot 2$.

b) Prime integers: An integer $p \in \mathbb{Z}$ is called prime, if $p \neq 0$ and $p \neq \pm 1$ and if the only divisors of p are the numbers ± 1 and $\pm p$. Thus, the prime numbers of algebra are: $\pm 2, \pm 3, \pm 5, \pm 7, \pm 11, \pm 13, \ldots$.

c) The algorithmic division: Given two integers of algebra, a and b, with $b \neq 0$, there exists a unique pair of integers $\boldsymbol{q} \in \mathbb{Z}$ and $\boldsymbol{r} \in \mathbb{N}_0$ such that

$$a = qb + r, \quad with \ \ 0 \leq r < |b| \qquad\qquad (5-1-1)$$

We note that, in the division in $\mathbb{Z}$, **the remainder r is a non negative integer**, smaller than the absolute value of the divisor b.

For example, the division of -29 by 6, gives a quotient $q = -5$ and a remainder $r = 1$, since:

$$29 = 4 \cdot 6 + 5 \Rightarrow -29 = (-4) \cdot 6 - 5 \Rightarrow -29 = (-4) \cdot 6 \underbrace{-6 + 1}_{-5} \Rightarrow$$

$$-29 = (-5) \cdot 6 + 1, \ i.e. \ q = -5 \ and \ r = 1$$

Other examples:

Division of -23 by -7: $-23 = 4 \cdot (-7) + 5$, quotient $q = 4$, remainder $r = 5$

Division of -12 by -17: $-12 = 1 \cdot (-17) + 5$, quotient $q = 1$, remainder $r = 5$.

Division of -60 by 18: $-60 = (-4) \cdot 18 + 12$, quotient $q = -4$, remainder $r = 12$.

d) The greatest common divisor: The greatest common divisor (G.C.D) of two or more integers of algebra, **is the greatest of the positive common divisors** of the given numbers, and can be found by replacing the given integers by their absolute values.

For example, the G.C.D of -8 and 12 is the number 4.

e) The least common multiple: The least common multiple (L.C.M) of two or more integers of algebra, **is the positive least common multiple** of the given numbers, and can be found by replacing the given integers by their absolute values.

For example, the L.C.M of -6 and 14 is the number 42.

f) Relatively prime integers: The definition and all the properties of the relatively prime integers of arithmetic, as stated in section 2-1, hold true for the integers of algebra.

Example 5-1-1: Let $n \in \mathbb{Z}$. Show that n can take one of the three forms: $n = 3k$ or, $n = 3k + 1$ or, $n = 3k - 1$.

Solution

If we divide n by 3, the possible remainders are: 0 or 1 or 2.

If $r = 0, n = 3k, (k \in \mathbb{Z})$, if $r = 1, n = 3k + 1, (k \in \mathbb{Z})$, while if $r = 2$, then $n = 3\lambda + 2 = 3\lambda + 3 - 1 = 3(\lambda + 1) - 1 = 3k - 1$, where $k = \lambda + 1 \in \mathbb{Z}$.

Example 5-1-2: Assuming that $n \in \mathbb{Z}$, consider the set A consisting of all the numbers of the form $2n + \frac{1}{2}$, and the set B consisting of all the numbers of the form $2n - \frac{1}{2}$. Show that the two sets A and B do not share any common elements, i.e. show that $A \cap B = \emptyset$.

Solution

Let $2k + \frac{1}{2}$ be an element of A, and $2\lambda - \frac{1}{2}$ be an element of the set B, with $k, \lambda \in \mathbb{Z}$. If the sets A and B had some common elements, then, there would be some $k \in \mathbb{Z}$ and $\lambda \in \mathbb{Z}$, such that:

$$2k + \frac{1}{2} = 2\lambda - \frac{1}{2}, \quad or, \quad 2(\lambda - k) = 1 \qquad (*)$$

which shows that 2 divides 1, and this is not possible. Therefore, the sets A and B do not have any common elements, and thus $A \cap B = \emptyset$.

PROBLEMS

5-1-1) Let $n \in \mathbb{Z}$. Show that n can take one of the three forms: $n = 7k$ or $n = 7k \pm 1$ or $n = 7k \pm 2$ or $n = 7k \pm 3$.

5-1-2) In example 5-1-2, consider the set D consisting of all the numbers of the form $(2n + 1)/2$. Show that $D = A \cup B$.

5-1-3) Find the quotient and the remainder of the following divisions:

$$\{-15 \div 7\}, \quad \{-23 \div (-5)\}, \quad \{-53 \div 4\}$$

(Ans: $q = -3, r = 6, \quad q = 5, r = 2, \quad q = -14, r = 3$).

5-1-4) Find the G.C.D and the L.C.M of the numbers 12 and -18.

(Ans: $(12, -18) = 6, \quad [12, -18] = 36$).

5-2) Theorems on the remainders

In this section we shall state some important theorems concerning the remainders of the division of two integers. These theorems, as we shall see, lead to some very useful and interesting applications.

Theorem 5-1: Let a, b and $k \neq 0$ be three algebraic integers, $(a, b, k \in \mathbb{Z})$. The two integers a and b leave the same remainder when divided by $k \neq 0$, if and only if k divides the difference $(a - b)$.

Proof: In general, if a and b are divided by k, we have (see eq. (5-1-1)):

$$a = qk + r_1 \qquad 0 \leq r_1 < |k|$$

$$\text{(5 - 2 - 1)}$$

$$b = Qk + r_2 \qquad 0 \leq r_2 < |k|$$

a) Assume that $r_1 = r_2 = r$. Then, $a = qk + r$, $b = Qk + r$, and subtracting the second from the first equation we find:

$$a - b = k(q - Q)$$

This equation shows that $(a - b)$ is a multiple of k, or, equivalently, k divides $(a - b)$, $(k \, / \, (a - b))$.

b) Conversely, let us assume that $k \, / \, (a - b)$. We shall show that the remainders r_1 and r_2 of the divisions $\{a \div k\}$ and $\{b \div k\}$, respectively, are equal.

Since $k \, / \, (a - b)$, by assumption, $a - b = nk$, with some $n \in \mathbb{Z}$. The first equation in (5-2-1) becomes, (since $a = b + nk$):

$$b + nk = qk + r_1, \quad or, \quad b = (q - n)k + r_1$$

and comparing with the second equation in (5-2-1), we find, $Q = q - n$ and $r_1 = r_2$.

For example, since $3 \, / \, (13 - 4)$, the numbers 13 and 4 leave the same remainder 1 when divided by 3. Indeed, $13 = 4 \cdot 3 + 1$ and $4 = 1 \cdot 3 + 1$. Also, since $4 \, / \, \big(11 - (-5)\big)$, the numbers 11 and -5 leave the same remainder 3 when divided by 4. Indeed, we find easily that $11 = 2 \cdot 4 + 3$ and $-5 = (-2) \cdot 4 + 3$.

Theorem 5-2: Let $a_1, a_2, \ldots, a_n \in \mathbb{Z}$ **and** $k \neq 0 \in \mathbb{Z}$. **Then, the remainder of the division of the sum** $(a_1 + a_2 + \cdots + a_n)$ **by** k **is not affected, if some or all the summands are replaced by the remainder of their division by** k. **In other words, if** r_1 **is the remainder of the division** $\{a_1 \div k\}$, r_2 **is the remainder of the division** $\{a_2 \div k\}$, **...,** r_n **is the remainder of the division** $\{a_n \div k\}$, **then:**

$$The\ remainder\ of\ the\ division\ \{(a_1 + a_2 + \cdots + a_n) \div k\}$$
$$= The\ remainder\ of\ the\ division\ \{(r_1 + r_2 + \cdots + r_n) \div k\}$$

Proof: From the equality of the algorithmic division we have:

$$a_1 = q_1 k + r_1, \qquad a_2 = q_2 k + r_2, \qquad \ldots\ldots \quad a_n = q_n k + r_n \qquad (*)$$

or, equivalently,

$$a_1 - r_1 = q_1 k, \qquad a_2 - r_2 = q_2 k, \qquad \ldots\ldots \quad a_n - r_n = q_n k \qquad (**)$$

Adding together these equalities we find,

$$(a_1 + a_2 + \cdots + a_n) - (r_1 + r_2 + \cdots + r_n) = (q_1 + q_2 + \cdots + q_n)k$$

where $(q_1 + q_2 + \cdots + q_n) \in \mathbb{Z}$.

Since k divides the difference $(a_1 + a_2 + \cdots + a_n) - (r_1 + r_2 + \cdots + r_n)$, the numbers $(a_1 + a_2 + \cdots + a_n)$ and $(r_1 + r_2 + \cdots + r_n)$ leave the same remainder when divided by k, according to theorem 5-1, and this completes the proof.

Theorem 5-3: Let $a_1, a_2, \ldots, a_n \in \mathbb{Z}$ **and** $k \neq 0 \in \mathbb{Z}$. **Then, the remainder of the division of the product** $(a_1 a_2 \ldots a_n)$ **by** k **is not affected, if some or all the factors are replaced by the remainder of their division by** k. **In other words, if** r_1 **is the remainder of the division** $\{a_1 \div k\}$, r_2 **is the remainder of the division** $\{a_2 \div k\}$, **...,** r_n **is the remainder of the division** $\{a_n \div k\}$, **then:**

$$The\ remainder\ of\ the\ division\ \{(a_1 a_2 \ldots a_n) \div k\}$$
$$= The\ remainder\ of\ the\ division\ \{(r_1 r_2 \ldots r_n) \div k\}$$

Proof: To illustrate the proof, let us begin with the first two numbers $a_1 = q_1 k + r_1$ and $a_2 = q_2 k + r_2$, whose product is

$$a_1 a_2 = (q_1 q_2 k + r_1 q_2 + r_2 q_1)k + r_1 r_2 \tag{*}$$

which implies that: $a_1 a_2 - r_1 r_2 = (q_1 q_2 k + r_1 q_2 + r_2 q_1)k$, with $(q_1 q_2 k + r_1 q_2 + r_2 q_1) \in \mathbb{Z}$. Since k divides the difference $(a_1 a_2 - r_1 r_2)$, it follows from theorem 5-1, that the remainder of the division $\{a_1 a_2 \div k\}$ is equal to the remainder of the division $\{r_1 r_2 \div k\}$.

Next, we consider the three numbers a_1, a_2 and a_3. From equation (*), we note that $a_1 a_2 = Ak + r_1 r_2$, where $A = (q_1 q_2 k + r_1 q_2 + r_2 q_1) \in \mathbb{Z}$. Then,

$$a_1 a_2 a_3 = (Ak + r_1 r_2)(q_3 k + r_3) = (Aq_3 k + r_1 r_2 q_3 + Ar_3)k + r_1 r_2 r_3$$

which implies that

$$a_1 a_2 a_3 - r_1 r_2 r_3 = (Aq_3 k + r_1 r_2 q_3 + Ar_3)k \tag{**}$$

where $(Aq_3 k + r_1 r_2 q_3 + Ar_3) \in \mathbb{Z}$.

Since k divides the difference $\{(a_1 a_2 a_3) - (r_1 r_2 r_3)\}$, the numbers $(a_1 a_2 a_3)$ and $(r_1 r_2 r_3)$ leave the same remainder when divided by k, according to theorem 5-1.

Similarly, working step by step, the theorem is proved for any number of factors.

Theorem 5-4: Let $a \in \mathbb{Z}, k \in \mathbb{Z}$ ($k \neq 0$), n be any natural number ($n \in \mathbb{N}$), and r be the remainder of the division $\{a \div k\}$. Then, the remainder of the division $\{a^n \div k\}$ is not affected if a is replaced by r, i.e.

$$\textit{The remainder of the division } \{a^n \div k\}$$
$$= \textit{The remainder of the division } \{r^n \div k\}$$

Proof: Follows readily from Theorem 5-3, if $a_1 = a_2 = \cdots = a_n = a$.

Comment: For economy in the notation, we shall denote by $R\{a \div k\}$ the remainder of the division of a by k, ($a \in \mathbb{Z}, k \neq 0 \in \mathbb{Z}$). For example, $R\{5 \div 3\} = 2, R\{-13 \div 5\} = 2$, etc.

Example 5-2-1: Find the remainder of the division of the number $a = 5 \cdot 17^2 + 7 \cdot 8^3$ by the number $k = 3$.

Solution

We note that: $R(5 \div 3) = 2, R(17 \div 3) = 2, R(7 \div 3) = 1, R(8 \div 3) = 2.$

Using theorems 5-2, 5-3 and 5-4 we have:

$$R\{(5 \cdot 17^2 + 7 \cdot 8^3) \div 3\} = R\{(2 \cdot 2^2 + 1 \cdot 2^3) \div 3\} = R\{16 \div 3\} = 1$$

Example 5-2-2: Find the remainder of the division $\{212^{35} \div 7\}$.

Solution

Since $212 = 30 \cdot 7 + 2$, the remainder of the division $\{212 \div 7\}$ is 2. According to theorem 5-4, we have:

$$R\{212^{35} \div 7\} = R\{2^{35} \div 7\} = R\{(2^3)^{11} \cdot 2^2 \div 7\} = R\{8^{11} \cdot 4 \div 7\}$$

and since the remainder of the division of 8 by 7 is 1,

$$R\{8^{11} \cdot 4 \div 7\} = R\{1^{11} \cdot 4 \div 7\} = R\{4 \div 7\} = 4$$

and finally, $R\{212^{35} \div 7\} = 4.$

Example 5-2-3: Find the remainder of the division of the number $a = 6563^{27} \cdot 2923^{19}$ by the number $k = 8$.

Solution

Since $6563 = 820 \cdot 8 + 3$, $R\{6563 \div 8\} = 3$, and since $2923 = 365 \cdot 8 + 3$ the remainder $R\{2923 \div 8\} = 3$. Applying theorems 5-3 and 5-4, we find:

$$R\{6563^{27} \cdot 2923^{19} \div 8\} = R\{3^{27} \cdot 3^{19} \div 8\} = R\{3^{46} \div 8\} = R\{(3^2)^{23} \div 8\}$$
$$= R\{9^{23} \div 8\} = R\{1^{23} \div 8\} = R\{1 \div 8\} = 1$$

Note that $R\{9^{23} \div 8\} = R\{1^{23} \div 8\}$, since the remainder of the division of 9 by 8 is 1.

Example 5-2-4: Show that 7 divides the number $a = 222^{555} + 555^{222}$.

Solution

It suffices to show that the remainder of the division of a by 7, is zero.

We note that $222 = 31 \cdot 7 + 5$, i.e. $R\{222 \div 7\} = 5$ and $555 = 79 \times 7 + 2$ i.e. $R\{555 \div 7\} = 2$. By virtue of theorems 5-2 and 5-4, we have:

$$R\{(222^{555} + 555^{222}) \div 7\} = R\{(5^{555} + 2^{222}) \div 7\}$$
$$= R\{((5^5)^{111} + (2^2)^{111}) \div 7\} \qquad (*)$$

However, $5^5 = 3125 = 446 \cdot 7 + 3$, and eq. (*) yields:

$$R\{((5^5)^{111} + (2^2)^{111}) \div 7\} = R\{(3^{111} + 4^{111}) \div 7\} \qquad (**)$$

Since 111 is an odd number,

$$3^{111} + 4^{111} = \underbrace{(3 + 4)}_{7} \cdot \underbrace{(3^{110} - 3^{109} \cdot 4 + \cdots - 3 \cdot 4^{109} + 4^{110})}_{Integer} \qquad (***)$$

Equation (***) results from the following identity, which holds true **for n odd positive integer:**

$$x^n + y^n = (x + y)(x^{n-1}y - x^{n-2}y^2 + \cdots - x^2 y^{n-2} + y^{n-1}) \qquad (****)$$

From eq. (***), it follows that $7 / (3^{111} + 4^{111})$, and therefore, the remainder $R\{(3^{111} + 4^{111}) \div 7\} = 0$, and consequently, 7 divides the number $(222^{555} + 555^{222})$ as well.

Example 5-2-5: If $a = 1^k + 2^k + 3^k + 4^k + 5^k + 6^k + 7^k + 8^k$, with k being a positive integer, what is the remainder of the division $\{a \div 5\}$?

Solution

Using theorems 5-2 and 5-4, we have:

$$R\{(1^k + 2^k + 3^k + 4^k + 5^k + 6^k + 7^k + 8^k) \div 5\} =$$
$$R\{(1^k + 2^k + 3^k + 4^k + 0 + 1^k + 2^k + 3^k) \div 5\}$$
$$= R\{(2 + 2^{k+1} + 2 \cdot 3^k + 4^k) \div 5\} \qquad (*)$$

a) **Assume that k is an odd integer: $k = 2n + 1, \ n = 0, 1, 2, 3, \ldots$**

Equation (*) becomes:

$$R\{(2 + 2^{2(n+1)} + 2 \cdot 3^{2n+1} + 4^{2n+1}) \div 5\}$$
$$= R\{(2 + 4^{n+1} + 6 \cdot 9^n + 4 \cdot 16^n) \div 5\}$$
$$= R\{(2 + 4^{n+1} + 1 \cdot 4^n + 4 \cdot 1^n) \div 5\}$$
$$= R\{(2 + 4^{n+1} + 4^n + 4) \div 5\} = R\{(6 + (4 + 1) \cdot 4^n) \div 5\}$$
$$= R\{(6 + 5 \cdot 4^n) \div 5\} = R\{(1 + 0 \cdot 4^n) \div 5\} = R\{1 \div 5\} = 1$$

(In this derivation we have used, repeatedly, theorems 5-2, 5-3 and 5-4).

b) Assume that k is an even integer: $k = 2n$, $n = 1, 2, 3, ...$

Equation (*) becomes:

$$R\{(2 + 2^{2n+1} + 2 \cdot 3^{2n} + 4^{2n}) \div 5\} = R\{(2 + 2 \cdot 4^n + 2 \cdot 9^n + 16^n) \div 5\}$$
$$= R\{(2 + 2 \cdot 4^n + 2 \cdot 4^n + 1^n) \div 5\}$$
$$= R\{(3 + 4^{n+1}) \div 5\} \qquad (**)$$

Now, in eq. (**), as it stands, we cannot apply theorems 5-2, 5-3 and 5-4, (since the remainders of the divisions $(3 \div 5)$ and $(4 \div 5)$ are 3 and 4, respectively).

The number n will be either even, or odd.

1) If $n = 2m$, (even), $k = 4m$, $m = 1, 2, 3, ...$, eq. (**) becomes:

$$R\{(3 + 4^{2m+1}) \div 5\} = R\{(3 + 4 \cdot 16^m) \div 5\} = R\{(3 + 4 \cdot 1^m) \div 5\}$$
$$= R\{(3 + 4) \div 5\} = R\{7 \div 5\} = 2 \qquad (***)$$

2) If $n = 2m + 1$, (odd), $k = 4m + 2$, $m = 0, 1, 2, ...$, eq. (**) becomes:

$$R\{(3 + 4^{2(m+1)}) \div 5\} = R\{(3 + 16^{m+1}) \div 5\} = R\{(3 + 1^{m+1}) \div 5\}$$
$$= R\{4 \div 5\} = 4 \qquad (****)$$

We may summarize our findings as follows:

$$R\{a \div 5\} = \begin{cases} 1 & when \ k \ is \ odd, & (k = 1, 3, 5, ...) \\ 2 & when \ k = 4m, & (k = 4, 8, 12, ...) \\ 4 & when \ k = 4m + 2, & (2, 6, 10, ...) \end{cases}$$

PROBLEMS

5-2-1) Find the remainder of the following divisions:

$$(5 \cdot 4^3 + 8 \cdot 7^{25}) \div 3, \quad (2 \cdot 6^{30} + 7 \cdot 11^{47}) \div 5, \quad (92^{45} + 2 \cdot 43^{79}) \div 7$$

(Ans: Remainders: $1, 4, 3$).

5-2-2) Show that 7 divides the number $a = 2222^{5555} + 5555^{2222}$.

Hint: See example 5-2-4.

5-2-3) Find the $R\{(2^{25} - 1) \div 7\}$, **(Ans:** 1).

5-2-4) Find the remainder of the divisions: $\{729^{13} \div 11\}$, $\{(19^{30} \cdot 25^{45}) \div 8\}$, **(Ans:** 5, 1).

5-2-5) Find the $R\{(1 \cdot 2 \cdot 3 \cdots 13 \cdot 14 \cdot 15 + 37) \div 13\}$, **(Ans:** 11).

5-3) Congruences

Congruences were introduced in mathematics by the great mathematician Carl Friedrich Gauss (1777 – 1855), in his efforts to simplify problems concerning divisibility of integers.

Let a, b be two integers, $(a, b \in \mathbb{Z})$ and m be a positive integer, $(m \in \mathbb{N})$. We say that a is congruent to b, modulo m, if the difference $(a - b)$ is a multiple of m, (or, the same, if m divides the difference $(a - b)$, i.e. if $m / (a - b))$, and we write:

$$a \equiv b \ (mod \ \ m) \tag{5 - 3 - 1}$$

The relation in eq. (5-3-1) is called "**congruence**" and the number m is called "**the modulus**" of the congruence.

By virtue of theorem 5-1, **we may say that $a \equiv b \ (mod \ m)$ when the two integers a and b leave the same remainder when divided by m**, (this is an equivalent definition of the congruence).

Equation (5-3-1) may be written as

$$a - b = mul. m, \ \ or, \ \ a = b + mul. m \tag{5 - 3 - 2}$$

If m does not divide the difference $(a - b)$ we say that a and b are incongruent mod. m, and write $a \not\equiv b \pmod{m}$.

The congruence $a \equiv 0 \pmod{m}$ implies that m / a.

The following relations are obvious:

1) $a \equiv a \pmod{m}$.

2) If $a \equiv b \pmod{m}$, then $b \equiv a \pmod{m}$.

3) If $a \equiv b \pmod{m}$ **and** $b \equiv c \pmod{m}$, then $a \equiv c \pmod{m}$.

For example: $7 \equiv 1 \pmod{3}$, since $3 / (7 - 1)$, $45 \equiv (-4) \pmod{7}$, since $7 / (45 - (-4))$, $-29 \equiv -4 \pmod{5}$, since $5 / (-29 - (-4))$, for any $k \in \mathbb{Z}$ it is true that $k \equiv k \pmod{m}$, since $k - k = 0 = mul.\, m$.

Also, note that:

$$every\ odd\ number \equiv 1 \pmod{2}$$

$$every\ even\ number \equiv 0 \pmod{2}$$

Properties of congruences:

1) $\left.\begin{matrix} a \equiv b \pmod{m} \\ c \equiv d \pmod{m} \end{matrix}\right\} \implies (a \pm c) \equiv (b \pm d) \pmod{m}$

Proof: $a \equiv b \pmod{m}$ implies that $m / (a - b)$ and $c \equiv d \pmod{m}$ implies that $m / (c - d)$. It follows that $m / \{(a - b) + (c - d)\}$, i.e. $m / \{(a + c) - (b + d)\}$, and this means that $(a + c) \equiv (b + d) \pmod{m}$. Similarly we show that $(a - c) \equiv (b - d) \pmod{m}$.

2) If $a \equiv b \pmod{m}$ and k is any integer, then: $a + k \equiv b + k \pmod{m}$.

It follows from property 1, since $k \equiv k \pmod{m}$

3) $\left.\begin{matrix} a \equiv b \pmod{m} \\ c \equiv d \pmod{m} \end{matrix}\right\} \implies ac = bd \pmod{m}$

Proof: The given congruences imply that: $a = b + mul.\,m$, (see eq. 5-3-2), and $c = d + mul.\,m$, and multiplying these two equalities we find,

$$ac = (b + mul.\,m)(d + mul.\,m) \Rightarrow$$

$$ac = bd + (mul.\,m)\cdot(b + d + mul.\,m) = bd + some\ multiple\ of\ m$$

and this means that $ac \equiv bd \pmod{m}$. This property may be extended to any number of congruences with the same modulus.

4) If $a \equiv b \pmod{m}$, then: $a^n \equiv b^n \pmod{m}$, where n is any positive integer.

It follows easily from property 3.

5) If $a \equiv b \pmod{m}$ and $k \neq 0$ is any integer, then: $ak \equiv bk \pmod{m}$.

It follows easily from property 3.

Properties 1 and 3 show that two congruences with the same modulus can be added, subtracted or multiplied, member by member, as if they were equations.

One property that differentiates congruences from equations is that of the cancelation. Common factors $\neq 0$, **cannot always be cancelled from both members of a congruence**, (while in equations this is always possible). For example, both members of the congruence

$$36 \equiv 6 \pmod{10}$$

are divisible by 6, but if we cancel the common factor 6, we get the congruence $6 \equiv 1 \pmod{10}$, which, obviously, is not correct, (since $10 \nmid 5$).

The next property shows that cancellation of a common factor from both members of a congruence is possible, **if the modulus is also divisible by the common factor**. For example, in the congruence $36 \equiv 6 \pmod{10}$, the number 2 is a common factor of both members and of the modulus as well. The given congruence implies the congruence $18 \equiv 3 \pmod{5}$, which is correct, since $5\ /\ (18 - 3)$.

6) $a \equiv b \pmod{m} \Leftrightarrow ak \equiv bk \pmod{km},\ \ k \neq 0$.

Proof: $a \equiv b \pmod{m}$ implies that $m / (a - b)$, and in turn, this implies that $km / k(a - b)$, for any integer $k \neq 0$, i.e. $ka \equiv kb \pmod{km}$.

Conversely, $ak \equiv bk \pmod{km}$ implies that $km / (ak - kb)$, i.e. $km / k(a - b)$, i.e. $m / (a - b)$, i.e. $a \equiv b \pmod{m}$.

Next property tells us that cancellation can, still, be used when the modulus is not divided by the common factor, **provided that the common factor is relatively prime with the modulus**.

For example, consider the congruence $28 \equiv 4 \pmod 3$. The number 4 is a common factor of both members, and is relatively prime to the modulus 3. We may thus cancel the common factor from both members and this leads to $7 \equiv 1 \pmod 3$, which is correct.

We now present and prove this important cancellation law.

7) $\left. \begin{cases} ak \equiv bk \pmod m \\ k \text{ and } m \text{ Relatively Prime} \end{cases} \right\} \Longleftrightarrow a \equiv b \pmod m$

Proof: $ak \equiv bk \pmod m$ implies that $m / (ak - bk)$, i.e. $m / k(a - b)$, and since, by hypothesis, k and m are relatively prime numbers, $m / (a - b)$, (Theorem 2-1, Euclid's theorem), i.e. $a \equiv b \pmod m$.

Conversely, $a \equiv b \pmod m$ implies that $m / (a - b)$, and if this is true, then, it is also true that $m / \big(k(a - b)\big)$, i.e. $m / (ka - kb)$, and this means that $ak \equiv bk \pmod m$.

Comment: Congruences provide an extremely powerful tool for the study of integers. Congruences not only facilitate the calculations and make the statement of theorems much easier to state, but also and most important, simplify the proofs of many theorems. This is illustrated in the examples that follow.

Example 5-3-1: Show that: $5^3 \equiv 5 \pmod{18}$, $5^{5k} \equiv 1 \pmod{11}$, $(k \in \mathbb{N})$.

Solution

a) $5^3 = 125 \Longrightarrow 5^3 - 5 = 120 = 5 \cdot 18 \Longrightarrow 5^3 \equiv 5 \pmod{18}$.

b) Since $5^5 = 3125 = 284 \cdot 11 + 1$, it follows that $5^5 \equiv 1 \ (mod \ 11)$, and then, according to the property (4), $(5^5)^k \equiv 1^k \ (mod \ 11)$, and this shows that $5^{5k} \equiv 1 \ (mod \ 11)$, for any positive integer k.

Example 5-3-2: If $b \equiv 3a \ (mod \ 5)$, show that $3a^2 + 8ab - 3b^2 \equiv 0 \ (mod \ 25)$.

Solution

Since $b \equiv 3a \ (mod \ 5)$, it follows that $b - 3a = 5k$, with k some integer, i.e. $b = 3a + 5k$. Then:

$$3a^2 + 8ab - 3b^2 = 3a^2 + 8a(3a + 5k) - 3(3a + 5k)^2$$
$$= 25 \cdot \underbrace{(-3k^2 - 2ak)}_{integer}$$

This shows that the number $25 \ / \ (3a^2 + 8ab - 3b^2)$, and therefore, $(3a^2 + 8ab - 3b^2) \equiv 0 \ (mod \ 25)$.

Example 5-3-3: Find the last digit of the successive powers of 7, and of the successive powers of 3.

Solution

a) The last digit of any integer a, is the remainder of its division by 10. For instance, assume: $a = 4837 = 4 \cdot 10^3 + 8 \cdot 10^2 + 3 \cdot 10 + 7$, or

$$a = (4 \cdot 10^2 + 8 \cdot 10^1 + 3) \cdot 10 + 7$$

and this clearly shows that the last digit of a, (number 7 in our case), is the remainder of the division $\{a \div 10\}$.

In general, if we call x the last digit of an integer a, then, $a = x + mul. \ 10$, i.e. $\boldsymbol{x \equiv a \ (mod \ 10)}$. Thus, $3 \equiv 5783 \ (mod \ 10)$, $2 \equiv 3982 \ (mod \ 10)$, etc.

b) The following congruences hold modulo 10:

$$7^1 \equiv 7, \ \ 7^2 \equiv 9, \ \ 7^3 \equiv 3, \ \ 7^4 \equiv 1, \ \ 7^5 \equiv 7$$

From this point on, the remainders are repeated in the same order. For example, from $7^1 \equiv 7$ and $7^5 \equiv 7$, it follows that $7^1 \cdot 7^5 \equiv 7 \cdot 7$, (from property 3), i.e. $7^6 \equiv 7^2 \equiv 9$, then, similarly, $7^7 \equiv 7^3 \equiv 3$, etc. In summary, the last digit of 7^6 is the same with the last digit of 7^2 (which is 9), the last digit of 7^7 is the same with the last digit of 7^3 (which is 3), etc. In general, for any $k \in \mathbb{N}$:

$$7^{4k} \equiv 1, \quad 7^{4k+1} \equiv 7, \quad 7^{4k+2} \equiv 9, \quad 7^{4k+3} \equiv 3, \quad (mod\ 10)$$

c) Working as in part (b) we find:

$$3^{4k} \equiv 1, \quad 3^{4k+1} \equiv 3, \quad 3^{4k+2} \equiv 9, \quad 3^{4k+3} \equiv 7, \quad (mod\ 10)$$

Example 5-3-4: Find the last digit of the number $a = 5473^{13} \cdot 4987^{21}$.

Solution

Let us first consider the number 5473. According to part (a), in ex. 5-3-3, we have:

$$3 \equiv 5473\ (mod\ 10) \Longrightarrow 3^{13} \equiv 5473^{13}\ (mod\ 10) \qquad (*)$$

From part (c) in ex. 5-3-3: $3 \equiv 3^{13} \equiv 5473^{13}\ (mod\ 10)$, i.e. the last digit of the number 5473^{13} is 3. Similarly,

$$7 \equiv 4987\ (mod\ 10) \Longrightarrow 7^{21} \equiv 4987^{21} \equiv 7(mod\ 10) \qquad (**)$$

(from part (b) in ex. 5-3-3), i.e. the last digit of 4987^{21} is 7.

Since the first factor of a ends in 3 and the second factor ends in 7, their product ends in 1, i.e. the last digit of a is 1.

Example 5-3-5: If p is any prime number ≥ 5, show that $p^2 \equiv 1\ (mod\ 24)$.

Solution

By hypothesis, $p \neq 2$ and $p \neq 3$. We want to show that the number $p^2 - 1$ is divisible by 24. It suffices to show that $3\ /\ (p^2 - 1)$ and $8\ /\ (p^2 - 1)$. If we prove this, then the number $p^2 - 1$ will be divisible by $24 = 3 \cdot 8$ as well, since 3 and 8 are relatively prime numbers, (Euclid's theorem).

The number $p^2 - 1 = (p - 1)(p + 1)$ is divisible by 8, since it is the product of two consecutive even numbers, the number $(p - 1)$ (even) and the number $(p + 1)$ (even).

If the number p is divided by 3, we will have: $p = 3q + r$, where $r = 0, 1, 2$ and q some positive integer. However, $r = 0$ **is excluded**, since in this case $p = 3q$ and p would be a composite number, which contradicts our hypothesis that p is prime. When $r = 1, p = 3q + 1, p^2 - 1 = 3 \cdot (3q^2 + 2q)$ and this shows that $3 / (p^2 - 1)$. Similarly, when $r = 2, p = 3q + 2$, and then, $p^2 - 1 = 3 \cdot (3q^2 + 4q + 1)$, i.e. $3 / (p^2 - 1)$.

We have thus shown that always, $3 / (p^2 - 1)$ and $8 / (p^2 - 1)$, and consequently, $3 \cdot 8 = 24 / (p^2 - 1)$.

Example 5-3-6: Show that the number $a^2 + b^2$, (a, b integers), is divisible by 7, if and only if, both a and b are divisible by 7.

Solution

Any integer a can be expressed as $a \equiv 0, \pm1, \pm2, \pm3 \pmod 7$. This is easy to show, since, as we know, $a = 7k + r$, where k is an integer and $r = 0$, or 1, or 2, or 3, or 4, or 5, or 6. If $r = 4$, then $a = 7k + 4 = 7k + 7 - 3$ or, $a = 7(k + 1) - 3$, i.e. $a \equiv -3 \pmod 7$, etc.

A similar expression holds for b, $b \equiv 0, \pm1, \pm2, \pm3 \pmod 7$.

By virtue of property (4) we have:

$$a^2 \equiv 0,1,4,9 \pmod 7, \quad or \quad a^2 \equiv 0,1,2,4 \pmod 7 \qquad (*)$$

$$b^2 \equiv 0,1,4,9 \pmod 7, \quad or \quad b^2 \equiv 0,1,2,4 \pmod 7 \qquad (**)$$

The number $a^2 + b^2$ is divisible by 7, if and only if, $a^2 + b^2 \equiv 0 \pmod 7$, and as it follows from equations (*) and (**), this can happen only if $a^2 \equiv 0 \pmod 7$ **and** $b^2 \equiv 0 \pmod 7$, i.e. only if $7 / a$ **and** $7 / b$.

Example 5-3-7: Fermat numbers were mentioned in section 3-4 (Historical remarks). We mentioned that the Fermat number $F_5 = 2^{2^5} + 1 = 2^{32} + 1$ is a composite number, since it is divided by 641. In this example we shall show that $641 \ / \ F_5$, without explicitly calculating F_5. The proof is based on the congrunces and their properties.

Solution

Since $641 = 5 \cdot 2^7 + 1 = 5 \cdot 2^7 - (-1)$, we have:

$$-1 \equiv 5 \cdot 2^7 \quad (mod \ 641) \tag{$*$}$$

From property 4, it follows: $(-1)^4 \equiv 5^4 \cdot (2^7)^4 (mod \ 641)$, i.e.

$$1 \equiv 5^4 \cdot 2^{28} \quad (mod \ 641) \tag{$**$}$$

On the other hand, we note that $641 = 625 + 16$, i.e. $641 = 5^4 + 2^4$, i.e.

$$5^4 \equiv -2^4 \quad (mod \ 641) \tag{$***$}$$

Multiplying eqs. ($**$) and ($***$) term wise, (by virtue of property 3), we find:

$$1 \cdot 5^4 \equiv (5^4 \cdot 2^{28}) \cdot (-2^4) \quad (mod \ 641), \quad or$$

$$5^4 \equiv -5^4 \cdot 2^{32} \quad (mod \ 641) \tag{$****$}$$

Since $5^4 = 625$ and 641 are relatively primes, the cancelation law holds, (property 7), and therefore, eq. ($****$) implies:

$$1 \equiv -2^{32} \ (mod \ 641), i.e. \ 2^{32} + 1 \equiv 0 \ (mod \ 641), i.e. \ F_5 \equiv 0 \ (mod \ 641)$$

and this shows that $641 \ / \ F_5$, i.e. F_5 is a composite number.

PROBLEMS

5-3-1) Show the following properties:

a) If $a \pm b \equiv c \ (mod \ m)$, then $a \equiv c \mp b \ (mod \ m)$.

b) If $a \equiv b \ (mod \ m)$, then $a \pm km \equiv b \ (mod \ m)$.

c) If $a \equiv b \ (mod \ m)$ and $d \ / \ m$, then $a \equiv b \ (mod \ d)$.

d) If $a \equiv b \ (mod \ m)$, then $(a, m) = (b, m)$, (recall (a, m) is the G.C.D of a and m, and similarly (b, m) is the G.C.D of b and m).

5-3-2) Show that from: $ab \equiv 0 \pmod{p}$, with p prime, it follows $a \equiv 0 \pmod{p}$ or $b \equiv 0 \pmod{p}$. Is this true in general if p is not a prime number?

5-3-3) Let $f(x) = a_n x^n + a_{n-1} x^{n-1} + \cdots + a_1 x + a_0$ be a polynomial with integer coefficients, $(a_i \in \mathbb{Z}, \ i = 0,1,2,\ldots,n)$. Using properties of congruences, show that if $k \equiv \lambda \pmod{m}$, then, $f(k) \equiv f(\lambda) \pmod{m}$.

5-3-4) Show that the congruence $a \equiv \pm 3 \pmod 8$ implies the congruence $a^2 \equiv 9 \pmod{16}$.

5-3-5) Show that $3^{4k+2} + 2 \cdot 4^{3k+1} \equiv 0 \pmod{17}$, $k \in \mathbb{N}$.

5-3-6) If $a \equiv b \pmod 2$ and $2 \,/\, (ab)$, show that $2 \,/\, a$ and $2 \,/\, b$.

5-3-7) If $k \in \mathbb{N}$, show that $23^k \equiv (-1)^k \pmod 4$ and $23^k \equiv (-1)^k \pmod 3$, and then show that $12 \,/\, \left(23^k - (-1)^k\right)$.

MISCELLANEOUS PROBLEMS

1) For what values of $k < 48$, $k \in \mathbb{N}$, the number $5k$ is divisible by 7 ?

(Ans: $k = 7, 14, 21, 28, 35, 42$).

2) If the numbers a and b are relatively prime, i.e. if $(a, b) = 1$, show that $(ab, a + b) = 1$ and $(a - b, ab) = 1$.

Hint: It suffices to show that ab and $a + b$ do not share any common **prime divisors**. If we assume that p is a common prime divisor of ab and $a + b$, then, since $p\ /\ ab$, it must divide at least one of the two numbers, say the number a, i.e. $p\ /\ a$. But, since by hypothesis, $p\ /\ (a + b)$ as well, it follows that $p\ /\ \{(a + b) - a\}$, i.e. $p\ /\ b$. In other words, p will be a common divisor of a and b. But a and b cannot have a common prime divisor, since by assumption, $(a, b) = 1$. Thus, the numbers ab and $a + b$ do not have common prime divisors, and consequently, they do not have any common composite divisors.

3) If $(a, b) = 1$ and $b \neq mul.\ 3$, show that $(3a + 2b, b) = 1$.

Hint: It suffices to prove that $3a + 2b$ and b do not have a common prime divisor p.

4) If a number a is the sum of the squares of two integers, show that $2a$ is also the sum of the squares of two integers.

Hint: If $a = n^2 + m^2$, then: $2a = 2n^2 + 2m^2 = n^2 + m^2 + 2nm + n^2 + m^2 - 2nm = (n + m)^2 + (n - m)^2$.

5) If $a, b, c \in \mathbb{N}$ and $a^2 + 2b = c^2$ show that: **a)** The numbers a and c are both even or both odd, and **b)** The number $a^2 + b$ is the sum of the squares of two integers.

6) Show that the numbers $5a + 6b$ and $a + b$ have the same common divisors with the numbers a and b, and consequently, $(5a + 6b, a + b) = (a, b)$, $(a, b \in \mathbb{N})$.

7) Find two integers a and b such that: $(a, b) = 6$ and $a + b = 30$.

(Ans: $a = 6, b = 24$, or, $a = 12, b = 18$).

Hint: $a = 6A, b = 6B$, where $a + b = 6(A + B) = 30$, i.e. $A + B = 5$.

8) Given that $(a, b, c) = 1$ show that $(a + 2b, b, c) = 1$.

Hint: It suffices to show that $a + 2b, b, c$ do not have any common prime divisors.

9) For what values of $k \in \mathbb{N}$, each one of the following numbers is prime? $17k$, $39(k^2 - 8)$, $97(k - 11)$.

(**Ans:** $k = 1, k = 3, k = 12$).

Hint: The number 17 is a prime number. If $k \neq 1$, then, $17k$ would be a composite number, which contradicts the hypothesis, ($17k$ is prime). Thus, k must necessarily by 1.

10) If $k \in \mathbb{N}$, show that $64/(7^{2k} + 16k - 1)$ and $42 / (k^7 - k)$.

Hint: a) If we call $a_k = 7^{2k} + 16k - 1$, it is easy to show that $a_{k+1} = 7^{2(k+1)} + 16(k + 1) - 1$, i.e. $a_{k+1} = 49a_k - 12 \cdot 64k + 64$. Notice that $a_1 = 64$. Then, $a_2 = 49 \cdot 64 - 12 \cdot 64 \cdot 1 + 64 = mul. 64$, i.e. $64 / a_2$. Then, $a_3 = 49 \cdot a_2 - 12 \cdot 64 \cdot 2 + 64 = mul. 64$, (since $64 / a_2$), etc. Thus, step by step, we show that $64/(7^{2k} + 16k - 1)$ for all $k \in \mathbb{N}$.

b) It suffices to show that $k^7 - k$ is divided by 2 and by 3 and by 7, since then, it will be divided by the product $2 \cdot 3 \cdot 7 = 42$, (since 2,3,7 are prime numbers, see theorem 2-8).

11) Find the L.C.M and the G.C.D of the numbers: $6, 9, 15, 18$.

(**Ans:** $(6,9,15,18) = 3$, $[6,9,15,18] = 90$).

12) Find two integers a and b such that: $(a, b) = 36$ and $[a, b] = 756$

(**Ans:** $a = 36, b = 756$, or, $a = 108, b = 252$).

Hint: If a and b are divided by their G.C.D 36, they become relatively prime. So, if we call $A = a/36$, $B = b/36$, then, $(A, B) = 1$. At the same time the L.C.M of a and b is also divided by 36, i.e. $[A, B] = 756/36 = 21$, and since A and B are relatively prime, the L.C.M $[A, B] = A \cdot B = 21$, i.e. $A = 1, B = 21$, or, $A = 3, B = 7$, etc.

13) If $(a, b, c) = 1$, show that: $(a + b + c, ab + bc + ca, abc) = 1$, $(a, b, c$ are positive integers).

Hint: It suffices to show that the three numbers $a + b + c$, $ab + bc + ca$ and abc do not have any common **prime** divisors, since then, they will not have any common composite divisors either.

14) If $a, b \in \mathbb{N}$ and $(a, b) = 1$ show that $\left(ab, a^k + b^k\right) = 1$, $(k = 1, 2, 3, ...)$.

Hint: The same as in problem 13.

15) If $(k - \lambda) / (ka + \lambda b)$ and $(k - \lambda) / (kb + \lambda a)$, show that $(k - \lambda) / (a + b)(k + \lambda)$, $(a, b, k, \lambda \in \mathbb{N})$, (assume $k - \lambda \neq 0$).

16) If $a = mul. b$, show that $a^3 - a^2 = mul. (b^2)$.

17) Find two natural numbers a and b such that: $a - b = 12$, $a \leq 100$ and $(a, b) = 12$.

(Ans: $12, 24$ or $36, 24$ or $48, 36$ or $60, 48$ or $72, 60$ or $84, 72$ or $96, 84$).

Hint: $a = 12A, b = 12B$, with $12A \leq 100$, i.e. $A = 1, 2, 3, 4, 5, 6, 7, 8$, and $12A - 12B = 12$, i.e. $A - B = 1$, etc.

18) Find the G.C.D and the L.C.M of the numbers 1236 and 147.

19) Find all the positive integers having: **a)** Two divisors only, **b)** Three divisors only,   **(Ans: a)** all prime integers p, **b)** All integers of the form p^2, with p prime).

20) Find $\tau(n)$ and $\sigma(n)$ for $n = 18900$.

21) Show that $10^{11} \equiv -1 \pmod{11}$.

22) Show that the difference of the squares of two prime numbers is a composite number, except the case $p_1 = 2, p_2 = 3$.

23) If $(a, b) = 1$, show that $3 \nmid (a^2 + b^2)$.

Hint: Since $(a, b) = 1$, none of the integers is divisible by 3, or, one number is divided by 3 and the other is not. In the first case, $a = 3k \pm 1$ and $b = 3\lambda \pm 1$. In the second case $a = 3k$ and $b = 3\lambda \pm 1$.

24) Show that $2304 / \left(k\, 7^{2(k+1)} - (k + 1)\, 7^{2k} + 1\right)$, $k = 1,2,3,4, \ldots$

Hint: If we call $a = k\, 7^{2(k+1)} - (k + 1)\, 7^{2k} + 1$, we have:

$$a = k7^{2k}(7^2 - 1) - (7^{2k} - 1) = k7^{2k}(7^2 - 1) - ((7^2)^k - 1)$$

$$a = k7^{2k}(7^2 - 1) - \left\{(7^2 - 1)(7^{2(k-1)} + 7^{2(k-2)} + \cdots + 7^2 + 1)\right\}$$

$$a = (7^2 - 1)\left\{k7^{2k} - 7^{2(k-1)} - 7^{2(k-2)} - \cdots - 7^2 - 1\right\}$$

$$a = (7^2 - 1)\left\{\left(7^{2k} - 7^{2(k-1)}\right) + \left(7^{2k} - 7^{2(k-2)}\right) + \cdots + \left(7^{2k} - 7^2\right) \right.$$
$$\left. + \left(7^{2k} - 1\right)\right\}$$

Notice that each parenthesis within the brackets is a multiple of $(7^2 - 1)$. For example, from the first parenthesis, we have,

$$7^{2k} - 7^{2(k-1)} = 7^{2(k-1)}(7^2 - 1)$$

and the same is true for all the other parentheses, and finally,

$$a = (7^2 - 1) \cdot (7^2 - 1) \cdot \{Integer\ number\} = 48^2 \cdot \{Integer\ number\}$$

This shows that $48^2 = 2304$ divides the number a.

25) Show that $10^{2k+1} \equiv -1 \pmod{11}$, $k \in \mathbb{N}$.

26) With $k \in \mathbb{N}$, show that the numbers $k(k + 1)/2$ and $2k + 1$ are relatively primes.

27) If $(a, b) = 1, (x, b) = 1, (y, a) = 1$ show that $(ax + by, ab) = 1$, $(a, b, x, y$ are positive integers).

Hint: It suffices to show that $ax + by$ and ab do not have any common, prime divisors, since then, they will not share any common composite divisors.

28) a) Let $a, b, k \in \mathbb{N}$. If the number a is a prime number and $ab = k^2$, show that $a \mathbin{/} b$, **b)** Show that the product of two unequal prime numbers cannot be the square of an integer number.

29) Show that $a^{k+4} \equiv a^k \pmod{10}$, $(a \in \mathbb{Z},\ k$ positive integer).

Hint: To show that $10 \mathbin{/} \left(a^{k+4} - a^k\right)$, it suffices to show that $\left(a^{k+4} - a^k\right)$ is divided by 2 and by 5. Note that
$$a^{k+4} - a^k = a^k(a^4 - 1) = a^k(a - 1)(a + 1)(a^2 + 1)$$

This number is always divided by 2, since between three consecutive integers, $a - 1, a, a + 1$, at least one is even. To show that it is divisible by 5, note that any $a \in \mathbb{Z}$ can be written as: $a = 5n$ or $5n \pm 1$ or $5n \pm 2$. It is easy to check that each one of these expressions for a, make $a^{k+4} - a^k$ a multiple of 5.

30) Show that $\left(2^k + 3^k, 2^{k+1} + 3^{k+1}\right) = 1$, with $k \in \mathbb{N}$.

Hint: It suffices to show that $2^k + 3^k$ and $2^{k+1} + 3^{k+1}$ do not have common, prime divisors. If we assume that a common prime divisor p exists, then p must also divide the number $2^{k+1} + 3^{k+1} - 2 \cdot \left(2^k + 3^k\right) = 3^k$, and the only prime number that can divide 3^k is $p = 3$. But then, p must also divide $\left(2^k + 3^k\right) - 3^k$, i.e. p must divide 2^k, i.e. $p = 2$, and thus we arrive at a contradiction.

31) For every integer $k \in \mathbb{N}$, show that the quotient of the division $\{k(k + 1) \div 2\}$, when divided by 3, gives a remainder $\neq 2$.

32) With $k, \lambda \in \mathbb{Z}$ and $k \neq \lambda$, show that the remainders of the divisions $\{k \div (k - \lambda)\}$ and $\{\lambda \div (k - \lambda)\}$ are equal, while their quotients differ by 1.

33) If each one of the integers $a, b, c \in \mathbb{Z}$ is not divisible by 3, then show that each one of the numbers $a^2 + b^2 + c^2$ and $a^2 - b^2$ is divisible by 3.

Hint: Since the numbers are not divisible by 3, we may set: $a = 3k \pm 1$, $b = 3\lambda \pm 1$, $c = 3m \pm 1$, with k, λ, m integers.

34) If a and b are integers not divisible by 3, show that $a^6 \equiv b^6 \pmod{9}$.

35) If a, b, m, n are integers such that $(10m + n) / (100a + b)$, show that $(10m + n) / (an^2 + bm^2)$.

Hint: If we call $c = 10m + n$, it follows from the hypothesis that $100a + b = c \cdot k$, $(c, k$ integers). Then, $n = c - 10m$, and $an^2 + bm^2 = a(c - 10m)^2 + bm^2 = ac^2 + 100am^2 - 20am + bm^2 = ac^2 - 20amc + (100a + b)m^2 = ac^2 - 20amc + ckm^2$, since $100a + b = c \cdot k$, and it follows that $an^2 + bm^2 = mul. c$, i.e. $c = (10m + n) / (an^2 + bm^2)$.

36) Find the remainder of the division $\{756^{43} \cdot 3987^{27} \div 11\}$.

(Ans: 7).

37) For every integer k, with $k \nmid 2$ and $k \nmid 3$, show that $k^2 + 23 \equiv 0 \ (mod \ 24)$.

38) Show that there are no integers a and b, such that: $3a^2 - b^2 = 1$.

Hint: Every integer b can be expressed as $b = 3k$, or, $b = 3\lambda \pm 1$.

39) With $k \in \mathbb{Z}$, show that: $10^{6k+2} + 10^{3k+1} + 1 \equiv 0 \ (mod \ 111)$.

40) For what numbers k is the number $a = 2^{2k} + 2^k + 1$ divisible by 7? $(k \in \mathbb{N})$,   **(Ans:** For $k \neq mul. 3$, the number a is divisible by 7).

Hint: The number a can be written as: 3λ, or $3\lambda + 1$, or $3\lambda + 2$.

41) Find all the integer solutions of the equation: $(x + 1)(y + 3) = 3xy$.

(Ans: $(x = 5, y = 2), (x = 1, y = 6), (x = 2, y = 3), (x = 0, y = -3), (x = -4, y = 1), (x = -1, y = 0))$.

Hint: The given equation reduces to: $(2x - 1)(2y - 3) = 9$. Since the unknowns x and y are integers, we must have: $2x - 1 = 9$ and $2y - 3 = 1$, **or,** $2x - 1 = -9$ and $2y - 3 = -1$, **or,** $2x - 1 = 3$ and $2y - 3 = 3$, **or,** $2x - 1 = -3$ and $2y - 3 = -3$, **or,** $2x - 1 = 1$ and $2y - 3 = 9$, **or,** $2x - 1 = -1$ and $2y - 3 = -9$.

42) Find the positive integers a of the form $a = 3p \cdot 2^n$, with p prime and $n \in \mathbb{N}$, having the property that: $\sigma(a) = 3a$.   **(Ans:** 120, 672).

Hint:

$$\sigma(a) = \frac{3^2 - 1}{3 - 1} \cdot \frac{p^2 - 1}{p - 1} \cdot \frac{2^{n+1} - 1}{2 - 1} = 3a = 3^2 p \cdot 2^n \implies$$

$$4 \cdot (p + 1) \cdot (2^{n+1} - 1) = 3^2 p \cdot 2^n \implies (p + 1) \cdot (2^{n+1} - 1) = 3^2 p \cdot 2^{n-2} \implies$$

$$(8 - p)2^{n-2} = p + 1$$

It follows that $p < 8$, i.e. the possible values of the prime p are: 7, 5, 3, 2. We find that when $p = 7, n = 2$, and when $p = 5, n = 3$. The other two prime numbers 3 and 2 do not give integer values for n, and therefore, are not accepted.

43) If $(a, b) = 1$, show that $(a \pm b, a^2 \pm ab + b^2) = 1, (a, b \in \mathbb{N})$.

Hint: It suffices to show that the two numbers do not have common, prime divisors.

44) Let $a_n = 4^n - 3n - 1, n = 1,2,3, ...$ **a)** Show that: $a_{n+1} = 4a_n + 9n$, and **b)** Show that: $4^n \equiv 3n + 1 \quad (mod\ 9)$.

45) Find the non negative integers x, y, z which satisfy the relations:
$x + 3y + 12z = 80,\ 0 \leq x < 3,\ 0 \leq y < 4$.

(Ans: $x = 2, y = 2, z = 6$).

Hint: $80 = 3 \cdot (y + 4z) + x$, and since $0 \leq x < 3$, it follows that x is the remainder of the division $\{80 \div 3\}$, i.e. $x = 2$. Then, $y + 4z = (80 - 2)/3$, i.e. $26 = 4z + y$, and since $0 \leq y < 4$, the number y is the remainder of the division $\{26 \div 4\}$, which is $y = 2$, etc.

www.ingramcontent.com/pod-product-compliance
Lightning Source LLC
Chambersburg PA
CBHW082221290526
45794CB00009B/3625